The Essential Guide to Passing The Transportation Civil PE Exam
Written in the form of Questions

160 CBT Questions Every PE Candidate Must Answer

Request latest Errata, or add yourself to our list for
future information about this book by sending an
email with the book title, or its ISBN, in the subject
line to:
Errata@PEessentialguides.com
or
Info@PEessentialguides.com

The Essential Guide to Passing The Transportation Civil PE Exam Written in the form of Questions

160 CBT Questions Every PE Candidate Must Answer

Jacob Petro
PhD, PMP, CEng, PE

PE Essential Guides
Hillsboro Beach, Florida

Report Errors For this Book

We are grateful to every reader who notifies us of possible errors. Your feedback allows us to improve the quality and accuracy of our products.

Report errata by sending an email to Errata@PEessentialguides.com

---

The Essential Guide to Passing The Transportation Civil PE Exam Written in the form of Questions: 160 CBT Questions Every PE Candidate Must Answer

1st Edition Print 3.2

© 2023 Petro Publications LLC. All rights reserved.

Published December 2023

All content is copyrighted by Petro Publications LLC and its owners. No part, either text or image, may be used for any purpose other than personal use. Reproduction, modification, storage in retrieval systems or retransmissions, in any form or by any means, electronic, mechanical, or otherwise, for reasons other than personal use, without prior permission from the publisher is strictly prohibited.

For written permissions contact: Permissions@PEessentialguides.com

For general inquiries contact: Info@PEessentialguides.com or PEessentialguides@outlook.com

Imprint name: PE Essential Guides

Company owning this imprint: Petro Publications LLC. Established in Florida, 2023.

ISBN: 979-8-9891857-3-3

Disclaimer

The information provided in this book is intended solely for educational and illustrative purposes. It is important to note that the technical information, examples, and illustrations presented in this book should not be directly copied or replicated in real engineering reports or any official documentation.

While there may be resemblances between the examples in this book and real structures or facilities, users must exercise caution and conduct comprehensive verification of all information before implementing it in any practical setting. The author and all affiliated parties explicitly disclaim any responsibility or liability arising from the misuse, misinterpretation, or misapplication of the information contained in this book.

Furthermore, it is essential to understand that this book does not constitute legal advice, nor can it be considered as evidence or exhibit in any court of law. It is not intended to replace professional judgment, and readers are encouraged to consult qualified experts or seek legal counsel for any specific legal or technical matters.

By accessing and utilizing the information in this book, readers acknowledge that they do so at their own risk and agree to hold the author and all affiliated parties harmless from any claims, damages, or losses resulting from the use or reliance upon the information provided herein.

"Strive for perfection in everything you do. Take the best that exists and make it better. When it does not exist, design it."

Sir Henry Royce

Preface

General Information about the Book

This book is designed to help civil engineers pass the NCEES exam with its 2024 updated specifications, which is a prerequisite for obtaining the professional engineering PE license in the United States 2024 onwards. This book is tailored to provide you with comprehensive knowledge, detailed examples, and step-by-step solutions with ample graphics that are directly related to the subjects covered by the NCEES exam.

In this book, you will find an extensive collection of civil engineering problems that are carefully selected to build your knowledge, skills, and ability to apply fundamental principles and advanced concepts in the field of civil engineering. These problems are accompanied by detailed explanations, diagrams, and equations to help you understand the underlying principles and solve the problems efficiently and accurately.

Whether you are a recent graduate, an experienced engineer, or a professional who wants to obtain the engineering license in the United States, this book will prepare you for the exam and equip you with the necessary tools to succeed.

The book is structured in a way such that it provides the reader with a comprehensive understanding of the core topics that are covered by the NCEES exam 2024 specifications – which have more in-depth focus compared to the previous versions of the same exam.

The book provides the reader with a full coverage and understanding for the NCEES relevant exam topics and possible question scenarios during a real test situation. If certain topics or methods were not covered in this book, the book method of presentation will ultimately guide you on how to find the solution you seek on your own, know where to find it, and provide the solution on a timely manner that saves you time during the real exam.

The questions in this book are neither easy nor difficult. They are constructive and creative in nature. They have been authored in a way to have you remember the core concept of the engineering topic you seek. They are designed to challenge engineers to think critically and apply their knowledge to exam and real-world scenarios. These questions require a deeper level of analysis and understanding than simple recall of information. They may involve multiple steps or require the engineer to consider different perspectives or solutions, while they can be difficult and require significant amount of effort and research to solve them.

Reasons I wrote this Book

I decided to author this book because I have a strong passion for engineering. I have a deep interest and understanding of civil engineering, and I wanted to share this knowledge and passion with others.

I am also very enthusiastic about engineering, and this has allowed me to explore various concepts and develop unique perspectives.

By writing this book I hope I can inspire others to pursue and improve on their career in engineering, help them pass the NCEES exam, improve on their skills and advance their knowledge in this field and provide them with the tools they need to succeed.

Lastly, the energy and enthusiasm I have, and I brought into this work is infectious and I wanted to channel this energy into this fascinating project and share it with others. I strongly believe this book will be a valuable resource for anyone interested in learning more about civil engineering.

Acknowledgement and Dedication

I would like to thank the readers of this book, who I hope will find it informative, engaging, and thought-provoking. It is my sincere hope that this book will inspire others to pursue their own passion, and that it will serve as a valuable resource for all those interested in the field of engineering.

Table of Contents

Preface vii
    General Information about the Book vii
    Reasons I wrote this Book vii
    Acknowledgement and Dedication viii

About the Author xi

Selected Nomenclature xiii

Introduction 1
    General Description of this Book 1
    Book Structure 1

Book Parts 1

About the Exam 2
    General information 2
    Dissecting the Exam 3

How to use this Book 3

Which References to Own 4

2024 Exam Specifications 4

Map of Problems Presented 7

Project Management 7
Traffic Engineering 7
Design & Geometry 8
Geotechnical & Pavement 9
Drainage 10

Problems & Solutions 11

Project Management 13
Traffic Engineering 33
Design & Geometry 83
Geotechnical & Pavement 133
Drainage 151

References 171

References' Key Chapters 174
   General 174
   Chapters per Reference 174
   HCM Manual Volume 1: Concepts (*) 175
   HCM Manual Volume 2: Uninterrupted Flow (*) 176
   HCM Manual Volume 3: Interrupted Flow (*) 178
   HCM Manual Volume 4: Applications (*) 181
   (*) Navigating the HCM Manual 181
   MUTCD Manual 182
   HSM Manual 183
   AASHTO's Green Book GDHS-7 185
   The Roadside Design Guide RSDG-4 187
   Hydraulic Design of Highway Culverts FHWA HIF-12-026 188
   Guide for the Planning, Design and Operation of Pedestrian Facilities GFP-2 188
   Mechanistic-Empirical Pavement Design Guide MEPDG-3 189
   Guide for Design of Pavement Structures GDPS-4-M 190

About the Author

Dr. Petro is a professional engineer and a business leader with over 20 years of experience in leading and growing engineering companies. Throughout his career, he worked with some of the most prestigious engineering firms. With a vast background in design and construction, he earned a reputation for delivering innovative and cutting-edge projects throughout his career.

Dr. Petro is a civil engineer, he holds a Doctorate degree, he has earned a Professional Engineering (PE) license as well as Chartered Engineer (CEng) certification. Additionally, Dr. Petro has earned a Project Management Professional (PMP) certification, further demonstrating his expertise in managing complex projects. Over the years, he successfully led and managed teams of engineers, designers, and other professionals, overseeing complex projects from conception to completion.

Throughout his career, Dr. Petro designed and delivered numerous innovative and interesting projects that have contributed significantly to various industries he has worked in. His passion for engineering and business has driven him to publish several papers and articles in industry-leading journals and magazines. His work has been recognized as state-of-the-art and has been referenced by many industry professionals.

As an international civil engineer who worked across the globe, Dr. Petro brings an interesting perspective to the table. He has a deep understanding of how civil facilities and structures work and how to optimize them for maximum efficiency and safety. His ability to communicate complex engineering concepts to both technical and non-technical stakeholders has been key to his success.

This page is intentionally left blank

Selected Nomenclature

A	Area, Amount of soil loss	ETC	Estimate to Complete
ADA	Americans with Disabilities	FF	Free Float
ADT	Average Daily Traffic	FFS	Free Flow Speed
AADT	Average Annual Daily Traffic	FHD	Free Haul Diagram
CC	Crash Cost	F.S.	Factor of Safety
CMF	Crash Modification Factor	GI	Group Index
ACWP	Actual Cost of Work Performed	GM	Gross Margin
ACWS	Actual Cost of Work Scheduled	GW	Well Graded Gravel
ATS	Average Travel Speed	GP	Poorly Graded Gravel
B/C	Benefit Cost Ratio	H	Height, Elevation, Depth
BCWP	Budget Cost of Work Performed	HOV	High Occupancy Vehicles
BCWS	Budget Cost of Work Scheduled	HSO	Horizontal Sightline Offset
BFFS	Base Free Flow Speed	HW	Headwater
BLOS	Bicycle Level of Service Score	HV	Heavy Vehicle
BP	Break Point	ID	Interchange Density
CAF	Capacity Adjustment Factor	ISD	Intersection Sight Distance
CBR	California Bearing Ratio	K	Proportion of AADT that occurs during peak hour
CPI	Cost Performance Index	L	Length
CRCP	Continuously Reinforced Concrete Pavement	LC	Lateral Clearance
D	Depth, Directionality Factor	LEF	Load Equivalency Factor
DDHV	Directional Design Hourly Volume	LEH	Limit of Economic Hall
		LF	Late Finish
DDI	Diverging Diamond Interchange	LL	Liquid Limit
E	Energy	LOS	Level Of Service
EAC	Estimate At Completion	LS	Late Start, drop in slope, or slope factor
EB	East Bound		
EF	Early Finish	MEV	Million Entering Vehicles
ES	Early Start	MHD	Mass Haul Diagram
ESAL	Equivalent Single Axle Load	ML	Managed Lane

N	Number of lanes	QALY	Quality Adjusted Life Years
NA	Not Applicable	R	Radius, Rainfall intensity index, Crash rate
NC	Normally Consolidated		
NE	North-East	R/A	Roundabout
NPV	Net Present Value	RSI	Relative Severity Index
NW	Non-Weaving segment	S	Speed, Degree of Saturation
O&M	Operation and Maintenance	SAF	Speed Adjustment Factor
OC	Over Consolidated	SE	South-East
OCC	Occupancy	SF	Safety Factor
OCR	Over Consolidation Ratio	SD	Saturated Surface Dry Aggregate
OH	Overhead	SSD	Stopping Sight Distance
OI	Operating Income	SL	Storage Length
P	Permitter, Conservation Factor	SN	Structure Number
PC	Point of Curve	ST	Station
PCC	Point of Compound Curve	SW	South-West
PCE	Passenger Car Equivalent	SW	Well Graded Sand
PDO	Property Damage Only	SP	Poorly Grade Sand
PFFS	Percentage of Free-Flow Speed	SPI	Schedule Performance Index
PHF	Peak Hour Factor	SUT	Single Unit Truck
PI	Plasticity Index	TF	Total Float
PI	Point of Intersection	TLC	Total Lateral Clearance
PL	Plastic Limit	TT	Tractor Trailer
PSD	Passing Sight Distance	TW	Tailwater
PSI	Present Serviceability Index	W	Width, Weaving segment, Intensity factor
PT	Point of Tangent		
PTSF	Percent Time Spent Following	V	Volume, Speed, Traffic Volume
PV	Present Value	VL	Average Vehicle Length
PVC	Point of Vertical Curve	VMT	Vehicle-Miles Traveled
PVI	Point of Vertical Intersection	VR	Volume Ratio
PVT	Point of Vertical Tangent	%OHP	% Occupied on-Highway Parking
Q	Flow	a	Acceleration

adj	Adjusted		*y*	Vertical distance
d	Depth		*z*	Elevation
dia	Diameter			
ft	Feet			
g	Gravitational acceleration			
h	Hour			
hr	Hour			
in	Inch			
ksi	Kilo pound per square inch			
ksf	Kilo pound per square foot			
kcf	Kilo pound per cubic foot			
lcph	Lane Change per Hour			
min	Minute, Minimum			
mph	Miles Per Hour			
n	Porosity, number of years, manning coefficient			
p	Pedestrian			
ped	Pedestrian			
pcph	Passenger Car per Hour			
pcphpl	Passenger Car per Hour Per Lane			
pcpmpl	Passenger Car per Mile per Lane			
psi	pound per square inch			
psf	pound per square foot			
pcf	pound per cubic foot			
s, sec	Second			
w	Width, Moisture Content			
veh	Vehicle			
vph	Vehicle per Hour			
x	Horizontal distance			

This page is intentionally left blank

Introduction

General Description of this Book

The Essential Guide to Passing The Transportation Civil PE Exam is a guide designed in the form of questions. It aims to achieve a comprehensive coverage for the transportation engineering material. It attempts to cover numerous exam scenarios with the use of questions.

Although the guide is designed with the use of 160 questions, some questions require two outcomes for a solution making them worth two questions not one. Those questions were intentionally designed this way so that a certain concept can be delivered which is intertwined with another, also, to ensure that the strategy of comprehensive coverage, which is the backbone strategy behind this book, is achieved. All in all, our aim behind authoring this book is to make it a one stop shop for passing the Transportation Civil PE exam.

The questions contained in this book are well thought out and presented in a certain manner. The structure of this book was made as presented to serve the purpose of its strategy – which is ultimately passing the PE exam.

The answers to the questions are detailed and they are well referenced. Some answers provide several methodologies in their presentation for the readers' benefit. Questions and answers are accompanied by graphics and detailed step-by-step explanations to help engineers understand concepts better. Not only this, the aim of providing such detailed answers is to help engineers understand all methods and apply them during the exam so they are better equipped with all possible exam scenarios.

Book Structure

The 2024 transportation exam specifications consist of ten knowledge areas. Those knowledge areas are presented in the relevant section of this book.

For ease of reference, and in order to ensure a clear, and a complete coverage for all possible question scenarios, those ten knowledge areas are grouped into five parts in this book, and they are as follows:

1) Project Management – 15 Questions
2) Traffic Engineering – 45 Questions
3) Design & Geometry – 60 Questions
4) Geotechnical & Pavement – 20 Questions
5) Drainage – 20 Questions

The number of questions in each part was carefully determined in a way that speaks to the expected number of questions in a real exam, also, they have been made as such ensuring good coverage for the intended material behind relevant knowledge areas for each part.

Book Parts

As described earlier, the book problems are arranged in five parts as presented in the Problems and Solutions Section. The coverage for each of the five parts is summarized in the following paragraphs:

> The *Project Management Part* covers topics that are mostly relevant to cost estimating and scheduling, economic analysis such as present worth and lifecycle cost, and other topics that are relevant to the project management knowledge area of the 2024 exam specifications.
>
> The *Traffic Engineering Part* focuses on the traffic engineering knowledge area of the NCEES 2024 specifications along

with traffic signal design, traffic control design and safety requirements.

The focus of this part is to improve the use and understanding of important manuals such as the AASHTO HCM manual, the MUTCD manual, the Guide for the Planning, Design and Operation of Pedestrian Facilities GPF, and the Highway Safety Manual HSM.

This part contains questions with examples on interrupted and uninterrupted flow, intersection capacity and traffic analysis, lane capacities, pedestrian traffic, traffic forecasts, trip generation, queuing diagrams, signal and signal design, and other relevant topics.

The *Design & Geometry Part* consists of questions from knowledge areas that have geometrical design requirements and cross-sectional design as well. For instance, topics with horizontal curve design, and vertical curve design, intersection geometry and roadside cross section design are all covered in this part.

The focus of this part is to improve the use and understanding of important guides and manuals such as the AASHTO's Green Book, the Roadside Design Guide along with other guides and manuals that could be the focus of other parts as well.

The *Geotechnical & Pavement Part* contains questions from the geotechnical and pavement knowledge area. Soil sampling, geotechnical evaluation, slope evaluation, consolidation and laboratory tests and many other relevant topics are cover in this part.

This part aims to improve the use and understanding of the NCEES Handbook, the AASHTO's Guide for Pavement Structures along with the Mechanistic-Empirical Pavement Design Guide.

The *Drainage Part* has questions from the drainage knowledge area. With topics on hydrology, hydraulics, runoffs, stormwater collection, culvert hydraulics and many other relevant questions.

This part aims to improve the use and understanding of the NCEES Handbook along with AASHTO's Hydraulic Design of Highway Culverts.

For ease of reference, knowledge areas covered in each of these parts are presented at each part's page break. Also, for better and quick understanding for topics covered in those parts, you can refer to the Map of Problems Presented section.

Although this book is divided into five distinct parts, it is important to recognize that these parts are not entirely separate from each other. In fact, and due to the nature of transportation engineering, these parts share common themes and ideas that overlap in various ways.

About the Exam
General information
The NCEES PE exam is a rigorous exam that is administered in two sessions, a morning session, and an afternoon session. The morning session is four hours long that used to focus on broader, foundational concepts in the engineering field along with a wide range of engineering problems compared to the afternoon session. This session is replaced in 2024 with the new exam specifications and its focus will be more in-depth topics to the transportation exam.

The afternoon session is also four hours long and is generally more focused on specific areas of expertise, in this case transportation engineering.

Both the morning and the afternoon sessions with the 2024 specifications carry a similar weight when it comes to importance and depth of coverage.

Both sessions consist of multiple-choice questions, point-and-click, drag-and-drop, or fill in the blank type. Candidates are typically required to demonstrate their ability to analyze and solve complex engineering problems during those sessions.

All in all, this exam is designed to test not only one's knowledge and technical skills, but also their ability to think critically and work under pressure.

Dissecting the Exam

The exam consists of 80 questions presented in two sessions. Candidates are given a total of 480 minutes to solve those questions. This means that each question is allotted an average time of six minutes. It is very important to keep in mind that some questions during the real exam will take only one minute to solve while others could take up to ten minutes to complete. To prepare for the exam, it is crucial to practice solving questions that have longer duration and that are more difficult, which is what this book aims to provide.

I would like to provide a reflection from my own experience sitting for the NCEES PE exam: During the exam, I was thoroughly prepared to confidently answer all questions within the allotted time, and even expected to complete the exam at a faster pace. However, I regretfully failed to study two specific chapters of a certain manual that I had assumed would have a lower probability of being included in the exam. To my dismay, two difficult questions emerged from these chapters, leaving me with no alternative but to study the relevant material from the codes and manuals provided eating away from the exam time. Consequently, I spent approximately 20-30 minutes ensuring that my answers were accurate. This unexpected event significantly impacted the remainder of my allotted exam time, and although I ultimately passed, it was an avoidable situation.

This experience is one of the reasons why I strived to provide a comprehensive coverage of all possible topics and scenarios in this book, such that exam candidates do not have to go through this experience.

How to use this Book

The questions presented in this book are designed with a mix of varying lengths. Some may only take a minute or two to answer, while others may require up to 10 or 15 minutes. This design has been done as such intentionally, as it reflects the format of the actual PE exam with more questions that are longer and more difficult than the exam.

By practicing with questions of varying lengths, candidates will be better prepared to manage their time during the exam. They will learn to quickly identify less complex questions and move through them efficiently, while also having the skills to tackle the more time-consuming questions effectively.

It is important to note that practicing only short questions, or questions that are six minutes long, may not be enough to fully prepare for the exam.

Furthermore, the variety of questions' lengths presented in this book helps keep one's mind engaged and challenged. It can be easy to become bored or disengaged when faced with a series of similar questions, but by mixing up the lengths, candidates will be forced to stay focused and adapt to the different types of questions.

Therefore, it is important to practice longer and more difficult questions in addition to the shorter ones. This will help develop one's ability to think critically, analyze complex problems, and apply their knowledge to solve them. It will also help build one's endurance and focus, which is critical for success on this exam.

Moreover, it is important to note that it is okay to spend more time on difficult questions or even shorter ones during the practice sessions. This will help identify weaknesses and areas of improvement.

As a final note, it is important to view this book as a textbook as it contains, not only straightforward questions and answers, but comprehensive explanations and guidance and wide coverage to the most complex exam problems with detailed elaborations of both questions and answers.

Which References to Own

The question at hand is whether owning references is necessary at this stage or not, and, if so, how many references are required, and which ones should be owned, and which ones can be spared.

Before delving more into this, I would like to emphasize that exam takers will have a continuous access to all references mentioned in the NCEES exam specifications – also mentioned and summarized at the end of this book. Because of this, familiarizing oneself with all those references prior to the exam is key.

The exam, and this book, is based on ten references, along with the NCEES PE handbook, all those references are mentioned, and each chapter summarized in the last section of this book, and they are all required to pass the exam.

Although all references are provided during the CBT test, it is not advisable to familiarize oneself with those references during the exam. Hence comes the following question:

Shall exam candidates own (or have access to) all references to study prior to the exam?

The simple answer is: Yes

Candidates can buy those references to guarantee an uninterrupted access or borrow them from a library or a colleague. Owning all references is beneficial for the freedom to study at one's own pace. It is worth noting that the AASHTO Store provides bundled offers for students or exam takers from time to time.

2024 Exam Specifications

Effective April 2024 the exam will focus more on the depth part with some general, but relevant, items from the planning and project management, drainage, and geotechnical knowledge areas.

This guide takes this change into account and the questions were authored with the 2024 exam specifications in mind.

The 2024 exam specifications are presented in the following two pages for ease of reference, also, portions of this specifications are presented at the Problems & Solutions relevant page breaks for each part denoting which areas are covered in that part.

SN	Knowledge Area	Expected Number of Questions
1	**Project Management** A. Quantity and cost estimating B. Project schedules (e.g., activity identification and sequencing) C. Economic analysis (e.g., present worth, lifecycle costs)	6-9
2	**Traffic Engineering (Capacity Analysis, Transportation Planning, and Safety Analysis)** A. Uninterrupted flow (e.g., level of service [LOS], capacity) B. Street segment interrupted flow (e.g., level of service [LOS], running time, travel speed) C. Intersection capacity (e.g., at grade, signalized, roundabout, interchange) D. Traffic analysis (e.g., volume studies, peak hour factor, speed studies, modal split, trip generation, traffic impact studies) E. Traffic safety analysis (e.g., conflict analysis, crash rates, collision diagrams) F. Nonmotorized facilities analysis (e.g., pedestrian, bicycle) G. Traffic forecasts and monitoring H. Highway safety analysis (e.g., crash modification factors, Highway Safety Manual)	
3	**Roadside and Cross-Section Design** A. Forgiving roadside concepts (e.g., clear zone, recoverable slopes, roadside obstacles) B. Barrier design (e.g., barrier types, end treatments, crash cushions) C. Cross-section elements (e.g., lane widths, shoulders, bike lane, sidewalks, retaining walls) D. Nonmotorized design considerations (e.g., shared-use paths, bicycle facilities, pedestrian facilities, ADA compliance, traffic-calming features)	10-15
4	**Horizontal Design** A. Basic circular curve elements (e.g., middle ordinate, length, chord definition, radius definition, centerline stationing) B. Sight distance considerations C. Superelevation (e.g., rate, transitions, method, components) D. Special horizontal curves (e.g., compound/reverse curves, curve widening, coordination with vertical geometry)	7-11
5	**Vertical Design** A. Vertical alignment (e.g., geometrics, vertical clearance) B. Stopping and passing sight distance (e.g., crest curve, sag curve)	8-12

6	**Intersection Geometry** A. Intersection sight distance B. Interchanges (e.g., freeway merge, entrance and exit design, horizontal design, vertical design) C. At-grade intersection layout, including roundabouts	7-11
7	**Traffic Signals** A. Traffic signal timing (e.g., clearance intervals, phasing, pedestrian crossing timing, railroad preemption) B. Traffic signal warrants C. Traffic signal design	5-8
8	**Traffic Control Design** A. Permanent signs and pavement markings B. Temporary traffic control	5-8
9	**Geotechnical and Pavement** A. Sampling, testing, evaluation, and soil stabilization techniques (e.g., soil classifications, subgrade resilient modulus, CBR, R-values, field tests, slope stability) B. Soil properties (e.g., strength, permeability, compressibility, phase relationships) C. Compaction, excavation, embankment, and mass balance D. Traffic characterization parameters, pavement design, and rehabilitation procedures (e.g., flexible and rigid pavement)	6-9
10	**Drainage** A. Hydrology, including runoff and water quality mitigation measures B. Hydraulics, including culvert and stormwater collection system design (e.g., inlet capacities, pipe flow, hydraulic energy dissipation, peak flow mitigation/detention, open-channel flow)	8-12

Map of Problems Presented

Project Management

Problem 1.1 Project Cost Analysis	**Problem 1.2** Project Activity Sequencing
Problem 1.3 Budgetary Cost Acceptable Range	**Problem 1.4** Benefit Cost Analysis
(⁂) **Problem 1.5** Net Present Value Analysis	**Problem 1.6** Resources Histogram
Problem 1.7 Mass Haul Diagram	**Problem 1.8** Volume of Excavation
Problem 1.9 Salvage Value	**Problem 1.10** Profit and Loss
Problem 1.11 Benefit Cost Ratio for Road Improvement	**Problem 1.12** Capitalized Costs
Problem 1.13 Rate of Concrete Pouring	**Problem 1.14** Bond Effective Interest
Problem 1.15 Depreciated Rate	

Traffic Engineering

Problem 2.1 Vehicles Average Spacing	**Problem 2.2** Freeway Lane Width
(⁂) **Problem 2.3** Changing Freeway Density	**Problem 2.4** Multilane Highway LOS
(⁂) **Problem 2.5** Managed Lane Volume	(⁂) **Problem 2.6** LOS for Two-Lane Highway
Problem 2.7 Bike Lane LOS	**Problem 2.8** Average Pedestrian Space
(⁂) **Problem 2.9** Crossing Difficulty Factor	**Problem 2.10** Bicycle Meeting Events
Problem 2.11 Off-Street Walkway	(⁂) **Problem 2.12** Transit Travel Speed
Problem 2.13 Rapid Transit Max Speed	**Problem 2.14** Lane Capacity
(⁂) **Problem 2.15** Adjusted Saturation Rate for a Permitted Right Turn	**Problem 2.16** Adjusted Saturation Rate for a Permitted Left Turn
Problem 2.17 Queuing Diagram	**Problem 2.18** Weaving Segment
(⁂) **Problem 2.19** Managed Lane Weaving Segment Intensity Factor	**Problem 2.20** Crosswalk Occupancy Time
Problem 2.21 Merging Segment Density	**Problem 2.22** Roundabout bypass Lane
Problem 2.23 Two-Way Stop-Controlled Intersection	**Problem 2.24** Dilemma Zone Elimination
Problem 2.25 Diverging Diamond Interchange DDI	**Problem 2.26** Traffic Signal Warrant

Problem 2.27 Interior Lane Closure	Problem 2.28 Chevron Sign Alignment
Problem 2.29 Signal Offset and Cycle Length	Problem 2.30 Coordinated Signal System
Problem 2.31 Uniform Delay for a Signalized Intersection Through Lane	Problem 2.32 Lane Reduction Signs and Road Marking
Problem 2.33 Crash Rate	Problem 2.34 Equivalent Property Damage Only (EPDO) Score
Problem 2.35 Roadway Sideslope Modification	Problem 2.36 Countermeasures Proposed
Problem 2.37 Trip Generation Model	Problem 2.38 Roadway Density
Problem 2.39 Peak Hour Factor	Problem 2.40 Transit Average Speed
Problem 2.41 Skid Marks Distance	Problem 2.42 85th Percentile Speed
Problem 2.43 Yellow Interval	Problem 2.44 Toll Sign Board
Problem 2.45 School Crossing Signal	

Design and Geometry

Problem 3.1 Basic Horizontal Curve	Problem 3.2 Basic Vertical Curve
Problem 3.3 Points on Vertical Curve	Problem 3.4 Double Horizontal Curve
Problem 3.5 Compound Horizontal Curve	Problem 3.6 Addition of a Compound Horizontal Curve
Problem 3.7 Superelevation and Road Cross Section – Spiral Curve	Problem 3.8 Basic Superelevation
Problem 3.9 Superelevation Transition	Problem 3.10 Superelevation and Road Cross Section – Straight Tangent
Problem 3.11 Reverse Horizontal Curve	Problem 3.12 Curve Widening
Problem 3.13 Horizontal Sight Distance	(⁂) Problem 3.14 Horizontal Curve Coordinate System (1)
(⁂) Problem 3.15 Horizontal Curve Coordinate System (2)	Problem 3.16 Spiral Curve Length
Problem 3.17 Vertical Curve Length	Problem 3.18 Escape Ramp
Problem 3.19 Vertical Curve Design with a Bridge on Top (1)	Problem 3.20 Vertical Curve Design with a Bridge on Top (2)
Problem 3.21 Crest Curve Slope	Problem 3.22 Crest Curve Design Speed

Problem 3.23 Crest Curve PSD	(⁂) **Problem 3.24** Crest Vertical Curve Passing a Fixed Point	**Problem 3.45** Highway Parallel Entrance	**Problem 3.46** Drainage Channel Cross Section
Problem 3.25 Lowest Point on Sag Curve	**Problem 3.26** Horizontal and Vertical Curve Coordination	**Problem 3.47** Runout Distance	**Problem 3.48** Ramp Clear-Zone
Problem 3.27 Parking Lane Considerations	**Problem 3.28** Local Road Maximum Grade	**Problem 3.49** Barrier Offset	**Problem 3.50** Length of Barrier
Problem 3.29 Collector Road Minimum Roadway Width	**Problem 3.30** Superelevation Treatment in a Divided Arterial Road	**Problem 3.51** Guardrail at Curb	**Problem 3.52** Slope End Treatments
Problem 3.31 Freeway Median Width	(⁂) **Problem 3.32** Intersection Sight Distance Case B	**Problem 3.53** Sand-Filled Barrels	**Problem 3.54** Self-Restoring Crash Cushions
(⁂) **Problem 3.33** Skewed Intersection Sight Distance Case C1	(⁂) **Problem 3.34** Skewed Intersection Sight Distance Case C2	**Problem 3.55** Obstacles Located at Lane Merge Locations	**Problem 3.56** Americans with Disabilities ADA Act for Curbs
Problem 3.35 Intersection Sight Distance Case F	**Problem 3.36** Preferable Island Size	**Problem 3.57** Retaining Walls	**Problem 3.58** Traffic Calming Features
Problem 3.37 Deceleration Lane	**Problem 3.38** Storage Length	**Problem 3.59** Bridge Walkway Provision for Individuals with Disabilities	**Problem 3.60** Provision for Narrow Walkways
Problem 3.39 Sight Distance for a Track Crossing	**Problem 3.40** Ramp Width		
Problem 3.41 Loop Ramp Design Speed and Superelevation	**Problem 3.42** A Two-Lane Exit		
Problem 3.43 Highway Taper Exit	**Problem 3.44** Highway Taper Entrance		

Geotechnical and Pavement

Problem 4.1 Soil Moisture Content	**Problem 4.2** Settlement in Clay
Problem 4.3 Effective Stress Over Time	**Problem 4.4** Soil Shear Strength

Problem 4.5	Problem 4.6
Subbase Stabilization	Soil Properties
Problem 4.7	**(⁂) Problem 4.8**
Soil Classification System (1)	Soil Properties
Problem 4.9	**Problem 4.10**
Soil Classification System (2)	Soil Permeability Testing
Problem 4.11	**Problem 4.12**
Slope Stability/ Slope Safety Factor	Optimum Moisture Content
Problem 4.13	**Problem 4.14**
Consolidation Settlement	Distresses in Flexible Pavements
Problem 4.15	**Problem 4.16**
Rigid Pavement Cement Properties	Concrete Mix Design
Problem 4.17	**Problem 4.18**
Soil Resistance Value	Problem Soil
Problem 4.19	**Problem 4.20**
Equivalent Single Axle Load ESAL for a Rigid Pavement	Base Layer Thickness

Drainage

Problem 5.1	Problem 5.2
Watershed Rainfall Depth	Precipitation Methods
Problem 5.3	**Problem 5.4**
Seawater Canal	Water Channel
(⁂) Problem 5.5	**Problem 5.6**
Water Discharge External Forces	Head Losses

Problem 5.7	(⁂) Problem 5.8
Unconfined Aquifer	Elevation of Water Surface in Reservoir
Problem 5.9	**Problem 5.10**
Retention Pond Sizing	Travel Time for Shallow Flow
Problem 5.11	**Problem 5.12**
Retention Basin Design	Shallow Flow
Problem 5.13	**Problem 5.14**
Soil Loss Prevention	Soil Erodibility
Problem 5.15	**Problem 5.16**
Flow in a Gutter	Culvert Inlet Headwater Elevation
Problem 5.17	**Problem 5.18**
Culvert Outlet Headwater Elevation	Hydraulic Jump
Problem 5.19	**Problem 5.20**
Energy Loss	Pond Water Quality

(⁂) Questions flagged like this either exceed typical exam length, or could be slightly more difficult than their counterparts, but they contain crucial concepts that are worth exploring. Those questions may also contain two concepts and are worth two questions not one. We decided to combine them in one question to deliver a certain concept, idea, or area of knowledge. For study and exam preparation, we highly recommend that you practice all questions regardless of level of difficulty or length. Overall, in this guide, we reduced the number of simple questions and focused on the slightly more challenging ones to improve understanding and enhance chances of success.

PROBLEMS & SOLUTIONS

- ❖ **Project Management**
- ❖ **Traffic Engineering**
- ❖ **Design & Geometry**
- ❖ **Geotechnical & Pavement**
- ❖ **Drainage**

Level of Difficulty (✦)

In the following sections, few questions may exceed typical exam length or may require multiple values for an answer. You may encounter similar length or level of difficulty in your exam session but not all questions are going to be that long or difficult. However, we have elected to design this book for an ultimate experience for the PE exam and we did not want to leave anything for chance.

The new exam format with an all-in depth focus, is not going to have straight forward questions, few questions are going to be of a challenging nature, and, in some instances, you may encounter some unpredictable questions. As a result, our intention in adding some challenging questions in this book is to ensure good coverage and to prepare you thoroughly to pass the exam.

For ease of reference, difficult questions, or questions that are slightly longer than average will be flagged with the symbol (✦). We encourage you to attempt all questions regardless of their length or level of difficulty due to the important concepts they discuss and the material they attempt to deliver.

PROJECT MANAGEMENT

Knowledge Areas Covered

SN	Knowledge Area
1	**Project Management** A. Quantity and cost estimating B. Project schedules (e.g., activity identification and sequencing) C. Economic analysis (e.g., present worth, lifecycle costs)

PART I
Project Management

PROBLEM 1.1 *Project Cost Analysis*
The below table summarizes four activities which belong to a larger program:

Activity	Pre-decessor	Budget (USD)	Actual Cost (USD)	Progress
(A) Excavate pit	-	10,000	4,850	85%
(B) Compact & prep pit to receive precast units	(A)	2,000	-	0%
(C) Pour, cure, prep & transport precast units	-	18,000	8,500	60%
(D) Installation and backfilling	(C)	10,000	-	0%
		100%	40,000	13,350

Activity	Wk1	Wk2	Wk3	Wk4	Wk5	Wk6	Wk7	Wk8
A	■	■						
B			■					
C		■	■	■	■	■	■	
D								■

At the end of week 3 the following statement best describes the project progress:

(A) Project is ahead of schedule and the estimate at completion is expected to be $27,668.

(B) Project is ahead of schedule and the estimate at completion is expected to be $34,050.

(C) Project is delayed as excavation works are behind schedule, based on cost performance, the project is expected to finish with a total cost of $27,668.

(D) Project is delayed as excavation works are behind schedule, based on cost performance, the project is expected to finish with a total cost of $34,050.

PROBLEM 1.2 *Project Activity Sequencing*
The below table represents project activities, durations in days, predecessors, and successors:

Activity	Duration	Predecessor	Successor
A	2	-	B, C
B	5	A	F, G
C	2	A	E, D
D	4	C	H
E	2	C	G
F	7	B	I
G	3	B, E	I
H	2	D	I
I	2	F, G, H	-

Assume work starts on a Monday, ignoring weekends and holidays, Total Float and Free Float for activity 'G' are:

(A) 5, 1

(B) 4, 4

(C) 4, 0

(D) 3, 3

PROBLEM 1.3 *Budgetary Cost Acceptable Range*
The city is conceptualizing a large infrastructure project with an expected cost of nearly $25 million. The city appointed a consultant to develop the feasibility and determine the budget for authorization from city council.

Assuming the project would in fact cost the city $25 million upon completion, what is an acceptable range of a budgetary estimate the consultant can provide the city with?

(A) $17.5 million to $37.5 million

(B) $20.0 million to $32.5 million

(C) $21.25 million to $30.0 million

(D) $20.0 million to $30.0 million

PROBLEM 1.4 *Benefit Cost Analysis*

The following roadway segment is 5 *miles* long and experiences an Average Annual Daily Traffic (*AADT*) of 80,000 vehicles per day:

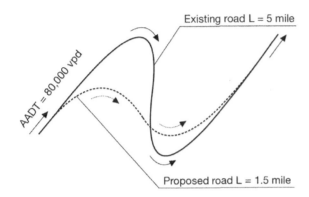

The dashed line is a proposed improvement/shortening of this segment which will cost $120.0 *million* to erect and is 3.5 *miles* shorter.

Considering the following factors:

- Average fuel cost along with operation & maintenance cost is $ 0.14 per vehicle per mile.
- Road maintenance cost is $22,000 per mile per year.
- Inflation rate is estimated at 2%

The Benefit/Cost ratio for this project over a period of 25 years is most nearly:

(A) 0.91

(B) 1.17

(C) 2.33

(D) 1.31

(⁂) PROBLEM 1.5 *Net Present Value Analysis*

One of the city's pumping stations require major mechanical rehabilitation for its pumps. The city has four options:

Option A: Replace old pumps with new pumps at a cost of $25,000 inclusive of labor. Yearly maintenance costs $3,000. The selected pumps to be replaced every 10 years.

Option B: Replace old pumps with new pumps at a cost of $27,500 inclusive of labor. Yearly maintenance costs $2,000. The selected pumps to be replaced every 10 years.

Option C: Replace old pumps with new pumps at a cost of $40,000 inclusive of labor. Yearly maintenance costs $1,500. The selected pumps to be replaced every 15 years.

Option D: Replace old pumps with new pumps at a cost of $47,500 inclusive of labor. Yearly maintenance costs $700. The selected pumps to be replaced every 15 years.

The remaining life of the civil structure is 30 years.

Average yearly inflation is 5% applied to capital and maintenance costs.

The city uses a discounted rate of 7% for their capital projects.

Based on the above, the cheapest and more feasible option amongst the four above would be:

(A) Option A

(B) Option B

(C) Option C

(D) Option D

(⁂) This question exceeds typical exam length, but it contains crucial concepts that are worth exploring.

PROBLEM 1.6 *Resources Histogram*

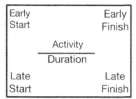

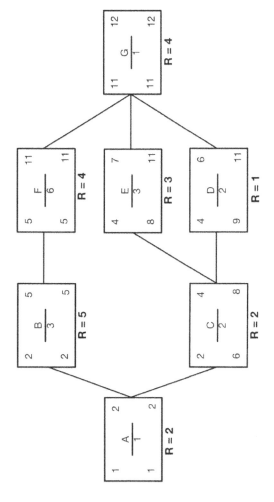

Assuming the activities in the above network diagram cannot be split, resources are of the same discipline, the below resources histogram represents the best possible leveled histogram (*):

(*) Activities start at the beginning of a day and finish at an end of a day – e.g., 'G' starts at the beginning of day 11 and ends at the end of day 11.

(A) Leveled Histogram A

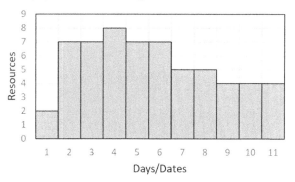

(B) Leveled Histogram B

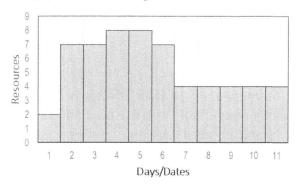

(C) Leveled Histogram C

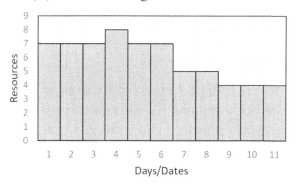

(D) Leveled Histogram D

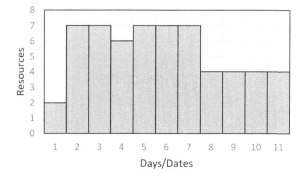

PROBLEM 1.7 *Mass Haul Diagram*

The below Mass Haul Diagram belongs to a highway with soil shrinkage factor of 14.5%

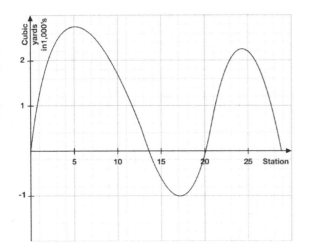

Considering a Free Haul Distance of 500 ft and a Free Haul Volume of 1,600 $yard^3$, wastage in $yard^3$ for this project is most nearly:

(A) 800

(B) 400

(C) 1,400

(D) 685

PROBLEM 1.8 *Volume of Excavation*

The below cut and fill diagram belongs to a highway project. The project's existing ground consists of soil with 20% shrinkage factor and 15% swell factor.

The diagram's negative y-axis represents cross section areas to be cut and the positive y-axis represents cross areas sections to be filled.

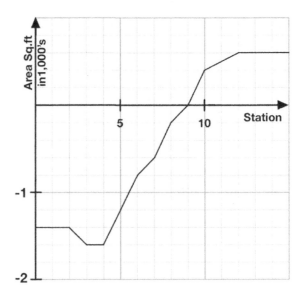

The project shall balance its cut and fill material and the rest goes to waste.

Considering dump trucks can haul up to 11 $yard^3$ per trip, the number of trips expected to haul waste outside the project is most nearly:

(A) 2,470

(B) 2,190

(C) 1,240

(D) 2,680

PROBLEM 1.9 *Salvage Value*

A company procured a fleet of high end, specially customized, old trucks for *five* of its employees to perform supervision services for a road construction contact $100 million in value at a cost of $55,000 per truck.

The company, which is a third-party contractor, priced its supervision services for this contract at 2.5% of the road construction fees.

The below is some of the key assumptions made by the company to support this agreement:

Item	Contract price
Average salary per employee	$80,000 / year$
Fringe benefits per employee	30%
Overhead cost	12.5% of company revenue
Allocations	7.5% of company revenue
Fuel & maintenance per truck	$0.25 / mile$
Expected miles driven for supervision purposes	$95\ mile/day/employee$
Working days per year (excluding vacation time)	$240\ days$
Contract duration	$2.5\ years$

In order to maintain an Operating Income OI of 15%, and considering those trucks are pretty much worn out after being driven over rough land for 2.5 $years$ in a row, they should be salvaged at an average minimum sales price of:

(A) $0 per truck

(B) $21,250 per truck

(C) $4,250 per truck

(D) $9,250 per truck

PROBLEM 1.10 *Profit and Loss*

You are a formwork/wood supplier who has been asked to provide material for one of the contractors to erect a building project. The material will cost you $100,000 to supply. The contractor asked you for a monthly payment plan over one year for cash flow reasons.

The monthly charge for the contractor that gets you to maintain an overall profit margin of 20% on all your expenses, knowing that bank charges a fixed interest of 6% per year, is most nearly:

(A) $10,600

(B) $10,952

(C) $10,332

(D) $10,720

PROBLEM 1.11 *Benefit Cost Ratio for Road Improvement*

A rural road segment is experiencing yearly crashes as presented in the following table:

Crash severity	Comprehensive Crash Unit Cost	Yearly Crashes before Widening	Yearly Crashes after Widening
Fatal (K)	$5,740,100	0	0
Disability (A)	$304,400	4	2
Evident (B)	$111,200	25	1
Possible (C)	$62,700	15	4
PDO (O)	$10,100	9	2

A road widening project with a cost of $15\ million$ is being considered and is expected to reduce yearly crashes as presented in the above table.

Considering a study period of 10 $years$ along with a rate of return of 5%, the Benefit Cost ratio for this project is expected to be:

(A) 2.7

(B) 2.1

(C) 2.5

(D) 0.51

PROBLEM 1.12 *Capitalized Costs*

The present net worth at a 4% interest rate for a project that has an initial cost of $1,000,000 and an Operation and Maintenance cost of $75,000 is most nearly:

(A) $2,875,000

(B) $1,875,000

(C) $1,750,000

(D) $2,171,650

PROBLEM 1.13 *Rate of Concrete Pouring*

The below top section is a plan view for a 14 in thick 11 ft high wall combination prepared for concrete placement for a certain facility.

The facility under consideration has eight of those wall combinations and concrete should be placed in sequence starting from wall combination 1 to wall combination 8.

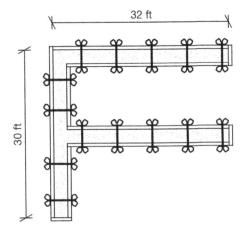

Concrete placement starts at 8:00 AM and must stop at 12:00 PM. Assuming concrete is pumped continuously at an average rate of 70 $yard^3/hr$ inclusive of lost time, a horizontal construction joint, if needed during the four-hour window, will be placed at:

(A) Wall combination 6 at a height of 0.5 ft

(B) Wall combination 6 at a height of 5 ft

(C) Wall combination 7 at a height of 5 ft

(D) Wall combination 7 at a height of 2.5 ft

PROBLEM 1.14 *Bond Effective Interest Rate*

A bond worth $200,000 that pays 7% yearly compounded interest realized every quarter for a year has an effective annual interest rate of:

(A) 7.2%

(B) 7.0%

(C) 7.5%

(D) 31%

PROBLEM 1.15 *Depreciated Rate*

A machinery was bought by a construction company for $125,000 and it was estimated that it would be salvaged in 10 years for $10,000 which represents the end of its life.

The rate of depreciation for this machinery is most nearly:

(A) 9%

(B) 10%

(C) 11%

(D) 12%

SOLUTION 1.1

Reference is made to the earned value management equations in *NCEES Handbook* based upon which the below tables are constructed:

Activity	I Duration (Weeks)	Budget (USD)	II Actual Cost (USD)	III Progress
A	2.0	10,000	4,850	85%
B	1.0	2,000	-	0%
C	6.0	18,000	8,500	60%
D	1.0	10,000	-	0%
	8.0	40,000	13,350	

IV Planned completion wk3	I × IV BCWS Planned Budget (USD)	II ACWP Actual Cost (USD)	I × III BCWP Earned Value (USD)
100%	10,000	4,850	8,500
100%	2,000	-	-
33.3%	6,000	8,500	10,800
0	-	-	-
	18,000	13,350	19,300

$CPI = BCWP/ACWP$
$= 19,300/13,350$
$= 1.445$

$SPI = BCWP/BCWS$
$= 19,300/18,000$
$= 1.07$

$ETC = (BAC - BCWP)/CPI$
$= (40,000 - 19,300)/1.445$
$= \$14,318$

$EAC = ACWP + ETC$
$= \$13,350 + \$14,318$
$= \$27,668$

The above indicates that the project is ahead of schedule with SPI >1.0 and cost performance is positive with CPI >1.0 and the estimate to completed is $27,668.

Correct Answer is (A)

SOLUTION 1.2

Using the critical path method, start with building an activity network. *Activity on nodes* was the preference in this solution with the following nomenclature and further steps:

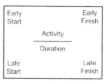

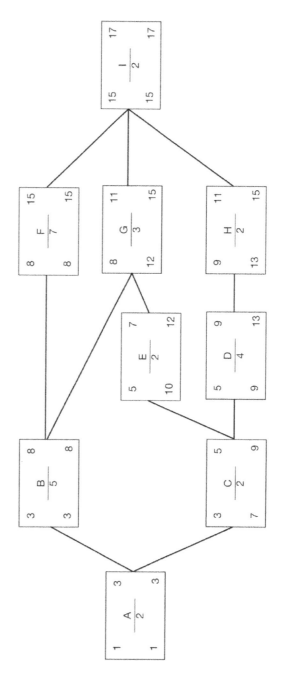

Step 1: A *forward pass* to determine early dates (Early Start ES and Early finish EF). When two activities feed forward into the same successor, the latest (longest) EF prevails.

Step 2: A *backward pass* to determine late dates (Late Start LS and Late finish LF). When two activities feed backward into the same predecessor, the earliest (shortest) LF prevails.

The following equations were applied to activity 'G':

$Total\ Float = LS - ES = 12 - 8 = 4\ days$

$Free\ Float = Earliest\ ES_{successor} - EF$
$= 15 - 11 = 4\ days$

Correct Answer is (B)

SOLUTION 1.3
This is a common request in real projects and is usually backed up with principles from the *Association for the Advancement of Cost Engineering* AACE.

The *Cost Estimate Classification Matrix* adopted by the AACE is found in the *NCEES Handbook version 2.0* Section 2.2.2. This matrix classifies estimates for Design Development, Budget Authorization and Feasibility as Class 3 Estimates.

The expected accuracy range for an AACE Class 3 estimate is:

L: − 5% to − 15%
(i.e., $21.25 million to $23.75 million)

H: + 10% to + 20%
(i.e., $27.5 million to **$30.0 million**)

The range would therefore be constituted from the lowest which is $21.25 *million* to the highest being $30.0 *million*.

Correct Answer is (C)

SOLUTION 1.4
To calculate the Benefit/Cost ratio for the road improvements project, we need to determine the total benefits and total costs over the 25 years period.

Costs:
Annual operation & maintenance and fuel costs are brought to their Present Values (PV) using Table 1.7.10 of the *NCEES Handbook version 2.0* with an interest rate of $i = 2\%$ and $n = 25\ years$ represented as $(P/A, 2\%, 25) = 19.5$. Cost is then calculated as follows:

(A) Annual cost before improvements is calculated as follow:

Road maintenance cost:
$= 5\ miles \times \$22,000 \times 19.5$
$= \$2,145,000\ (\$2.145\ million)$

The estimate for the annual Vehicle-Miles Traveled (VMT) is calculated using the *NCEES Handbook* Section 5.1.3.2:

$VMT_{365} = AADT \times L \times 365$
$= 80,000 \times 5 \times 365$
$= 146 \times 10^6\ miles\ per\ year$

Vehicle fuel and O&M cost:
$= 146 \times 10^6\ miles \times \$0.14 \times 19.5$
$= \$398.58\ million$

Total fuel and vehicle O&M cost $(TC_{O\&M\ before})$:
$= \$398.58\ million + \$2.145\ million$
$= \$400.73\ million$

(B) Annual cost after improvements is calculated as follow:

Road maintenance cost:
$= 1.5 \; miles \times \$\,22{,}000 \times 19.5$
$= \$\,643{,}500 \; (\$\,0.64 \; million)$

The estimate for the annual Vehicle-Miles Traveled (VMT) is calculated using the *NCEES Handbook* Section 5.1.3.2:

$VMT_{365} = AADT \times L \times 365$
$= 80{,}000 \times 1.5 \times 365$
$= 43.8 \times 10^6 \; miles \; per \; year$

Vehicle fuel and O&M cost:
$= 43.8 \times 10^6 \; miles \times \$\,0.14 \times 19.5$
$= \$\,119.574 \; million$

Total fuel and vehicle O&M cost ($TC_{O\&M \; after}$):
$= \$\,119.57 \; million + \$\,0.64 \; million$
$= \$\,120.21 \; million$

Total Cost after improvement (TC_{after}):
$= \$\,120.21 \; million + \$\,120.00 \; million$
$= \$\,240.21 \; million$

Benefits:
The benefits arise from the cost savings due to reduced fuel and O&M for vehicles, along with the maintenance cost for the improved road segment over the specified duration of 25 years.

$= TC_{O\&M \; before} - TC_{O\&M \; after}$
$= \$\,400.73 \; million - \$\,120.21 \; million$
$= \$\,280.52 \; million$

Benefit/Cost ratio:

B/C is calculated by dividing the total benefits (savings) by the total cost:

$B/C = \dfrac{Savings}{Total \; Cost(TC_{after})}$

$= \dfrac{\$\,280.52 \; million}{\$\,240.21 \; million}$

$= 1.17$

A B/C of 1.17 suggests that for every dollar invested in the road improvement, you would receive approximately $\$\,1.17$ in benefits over the 25 years period. Typically, a B/C greater than '1.0' is considered economically viable, as it indicates that the benefits outweigh the costs (*).

Correct Answer is (B)

(*) It is essential to consider other factors in the Benefit Cost analysis such as potential future savings, environmental impacts, improved safety, and additional intangible benefits that may not be captured in this simple analysis. A more comprehensive evaluation in real life examples may be necessary to make well-informed decisions.

SOLUTION 1.5
Although this seems a slightly longer question than an exam question would be, it provides an explanation of the concept of NPV and options analysis for a real, and frequent, engineering problem.

One of the means of performing options analysis is the *Net Present Value* (NPV) method. With this, future costs shall be determined with using average inflation. After that, NPV is calculated per option using the given *discounted rate* as *interest rate*.

Step 1: Calculate inflation and inflation factors as follows:

Step 2: Perform NPV analysis as follows:

Option A

Event	Year	n	Cost (I)	Inflation Symbol	Inflation Factor (II)		Inflated Cost I x II	NPV Symbol	NPV Factor	NPV
New pumps	0		$25,000	-	1.0000		$25,000.00	NA	1.0000	$25,000.00
New pumps	10	10	$25,000	(F/P, 5%, 10)	1.6289		$40,722.50	(P/F, 7%, 10)	0.5083	$20,699.25
New pumps	20	20	$25,000	(F/P, 5%, 20)	2.6533		$66,332.50	(P/F, 7%, 20)	0.2584	$17,140.32
Maintenance	1→9 & 10→19 & 20→30	28	$3,000	(F/A, 5%, 30) minus 2 x (F/P, 5%, 30)	66.4388 minus 2 x 4.3219		$173,385.00	(P/F, 7%, 30)	0.1314	$22,782.79
										$85,622.35

Option B

Event	Year	n	Cost (I)	Inflation Symbol	Inflation Factor (II)		Inflated Cost I x II	NPV Symbol	NPV Factor	NPV
New pumps	0		$27,500	-	1.0000		$27,500.00	NA	1.0000	$27,500.00
New pumps	10	10	$27,500	(F/P, 5%, 10)	1.6289		$44,794.75	(P/F, 7%, 10)	0.5083	$22,769.17
New pumps	20	20	$27,500	(F/P, 5%, 20)	2.6533		$72,965.75	(P/F, 7%, 20)	0.2584	$18,854.35
Maintenance	1→9 & 10→19 & 20→30	28	$2,000	(F/A, 5%, 30) minus 2 x (F/P, 5%, 30)	66.4388 minus 2 x 4.3219		$115,950.00	(P/F, 7%, 30)	0.1314	$15,235.83
										$84,359.35

Option C

Event	Year	n	Cost (I)	Inflation Symbol	Inflation Factor (II)		Inflated Cost I x II	NPV Symbol	NPV Factor	NPV
New pumps	0		$40,000	-	1.0000		$40,000.00	NA	1.0000	$40,000.00
New pumps	15	15	$40,000	(F/P, 5%, 15)	2.0789		$83,156.00	(P/F, 7%, 15)	0.3624	$30,135.73
Maintenance	1→14 & 15→30	29	$1,500	(F/A, 5%, 30) minus 1 x (F/P, 5%, 30)	66.4388 minus 1 x 4.3219		$93,175.35	(P/F, 7%, 30)	0.1314	$12,243.24
										$82,378.98

Option D

Event	Year	n	Cost (I)	Inflation Symbol	Inflation Factor (II)		Inflated Cost I x II	NPV Symbol	NPV Factor	NPV
New pumps	0		$47,500	-	1.0000		$47,500.00	NA	1.0000	$47,500.00
New pumps	15	15	$47,500	(F/P, 5%, 15)	2.0789		$98,747.75	(P/F, 7%, 15)	0.3624	$35,786.18
Maintenance	1→14 & 15→30	29	$700	(F/A, 5%, 30) minus 1 x (F/P, 5%, 30)	66.4388 minus 1 x 4.3219		$43,481.83	(P/F, 7%, 30)	0.1314	$5,713.51
										$88,999.70

Option (A) explained:
- Year zero: new pumps to be procured at a cost of $25k$. The $25k$ is the present value of this initial engineering decision and hence no factor is applied on it and this cost shall be taken to the finish line (i.e., NPV) as is.

- Year 10: new pumps to be procured at a cost of $25k$ *now*. In 10 years however, the cost of those pumps is expected to be $(F/P, 5\%, 10)$ more – i.e.,
$\$25,000 \times 1.6289 = \$40,722.5$

The $\$40,722.5$ is brought back to year zero using the city discounted rate (i.e., the best estimate for their return on investment on capital projects) using $(P/F, 7\%, 10)$ – i.e.,
$\$40,722.5 \times 0.5083 = \$20,699.25$

- Year 20: new pumps to be procured at a cost of $25k$ *now* as given in the question. In 20 years however, the cost of those pumps is expected to be $(F/P, 5\%, 20)$ more – i.e., $\$66,332.5$. This amount is brought back to year zero using $(P/F, 7\%, 20)$ as:
$(25,000 \times 2.6533) \times 0.2584 = 17,140.3$

- Years 1 to 30: yearly maintenance cost of $\$3,000$. With inflation, this cost is brought forward to year 30 in one lump with the factor $(F/A, 5\%, 30)$ as:
$\$3,000 \times 66.4388 = \$199,316.4$

There are two years however during when maintenance costs are not required which is when pumps are installed at year 10 and at year 20. Those two years' payments are deducted in year 30 with $2 \times (F/P, 5\%, 30)$ as:
$2 \times 4.3219 \times \$3,000 = \$25,931.4$

→ $\$199,319.4 - \$25,931.4 = \$173,385.0$

The overall net maintenance resultant in year 30 is therefore expected to be $\$173,385.0$. This amount is discounted (i.e., brought back to year zero) with the factor $(P/F, 7\%, 30)$ to:
$\$173,385.0 \times 0.1314 = \$22,782.79$

The total NPV for option A in this case is:

$\$25,000 + \$20,699.25 + \$17,140.32 +$
$\$22,782.79 = \$85,622.36$

A similar process is applied to all the given options and option C was the most feasible.

Correct Answer is (C)

SOLUTION 1.6
It is important to understand which activities can change their start and finish dates so they can be used for resources leveling. Those activities are the ones with positive free float. Free float can be determined using the following equation:

$Free\ Float = Earliest\ ES_{successor} - EF$

$FF_D = 11 - 6 = 5\ days$

$FF_E = 11 - 7 = 4\ days$

The following table summarizes the network diagram along with the available Free Floats:

Activity	Duration	Resources	ES	EF	FF
A	1	2	1	2	
B	3	5	2	5	
C	2	2	2	4	
D	2	1	4	6	5
E	3	3	4	7	4
F	6	4	5	11	
G	1	4	11	12	

This table is then converted into the base Gantt Chart shown below, out of which a resources histogram is drawn by assigning resources to the bar chart and summing them up in the histogram that follows.

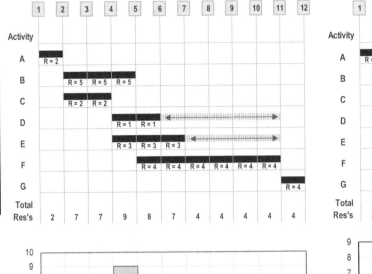

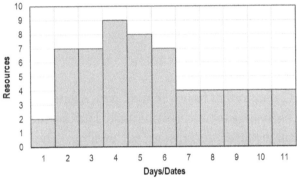

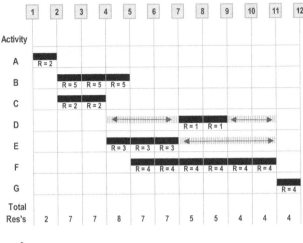

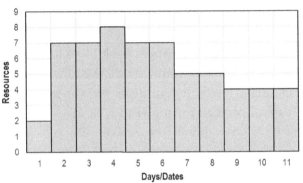

As pointed out in the above Gantt Chart, activities D and E can move freely as indicated by the two-sided arrows for *five* and *four* days respectively without affecting the completion date of the project.

In this case, the solution becomes a matter of trial and error. Activities D and E can move horizontally within their float to match the best available option from the four options provided in this question.

Based on this, the following Gantt chart was established by sliding activity D *three* days to the right which could provide the most leveled usage for those resources, or at least to match one of the histograms provided in the question.

The rest of the diagrams are simply incorrect.

Correct Answer is (A)

SOLUTION 1.7

The Free Haul Distance (FHD) has been given in the problem as $500\,ft$. The FHD is the distance below which earthmoving is considered part of the contract and contractor cannot claim for extras for overhauling.

To identify stations which fall within the FHD, a $500\,ft$ *to-scale* horizonal line is drawn and fit in position to intersect close to the peaks and troughs of the Mass Haul Diagram (MHD) curves as shown in the figure below. The y-axis generated values from the FHD intersection with the MHD curves represent the quantity which will be hauled as part of the contract price with no extras.

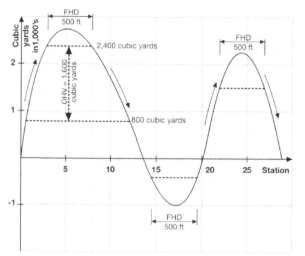

The Over Haul Volume (OVH) is the volume beyond which earthmoving can be claimed as extra by the contractor. OVH is the vertical distance from the FHD towards the x-axis (whether upwards or downwards) to an imaginary horizontal line that intersects with the two sides of the semi parabolic curves.

In this case, and as given in the question, the $OHV = 1,600 \; yard^3$ measured as a vertical distance from the FHD down to an intersection of $800 \; yard^3$ for the first MHD curve. This does not apply to the second or the third curves because there isn't enough vertical distance to establish such ordinates.

The vertical ordinates generated form this intersection represent either: (1) the waste material when the curve is moving upwards (i.e., cut sections which is more economical to dump outside the project), or (2) the borrow material when the curve is moving downwards (i.e., fill sections which is more economical to supply its material from outside the project). The horizonal intersection is hence called the Limit of Economic Hall (LEH). Shrinkage does not apply in the cut section therefore, wastage in this case will only be $800 \; yard^3$ (*).

Correct Answer is (A)

(*) To put things into perspective for this question, and as concluded from the MHD, the material that should go to waste is located between stations $'0 + 00'$ and $'0 + 70'$. If we assume this is a $60 \; ft$ wide highway construction project, the height of this cut section that should go to waste is nearly:

$$\frac{800 \; yard^3 \times 27 \frac{ft^3}{yard^3}}{60 \; ft \times 70 \; ft} \approx 5 \; feet$$

SOLUTION 1.8

The average area method is used to measure the volume for each two consecutive cross-sectional areas where the distance between consecutive areas is a station $= 100 \; ft$:

$$V = L \left(\frac{A_1 + A_2}{2} \right)$$

The following two tables represent cross-sections measured at each station along with the volume between every two consecutive stations.

Bank (undisturbed) cut volume $V_{B,cut} = -950,000 \; ft^3$ is converted to lose volume $V_{L,cut}$. Fill requirements $V_{B,fill}$ is also converted into loose volume $V_{L,fill}$. The balance of both goes to waste.

$$V_{L,cut} = \left(1 + \frac{S_w}{100}\right) V_{B,cut}$$

$$= \left(1 + \frac{15}{100}\right)(-950,000)$$

$$= -1,092,500 \; ft^3$$

$$V_{L,fill} = \left(1 + \frac{S_h}{100}\right) V_{B,fill}$$

$$= \left(1 + \frac{20}{100}\right)(300,000)$$

$$= 360,000 \; ft^3$$

Station	Cut Area ft^2	Fill Area ft^2
0+00	− 1,400	
1+00	− 1,400	
2+00	− 1,400	
3+00	− 1,600	
4+00	− 1,600	
5+00	−1,200	
6+00	− 800	
7+00	− 600	
8+00	− 200	
9+00	0	0
10+00		400
11+00		500
12+00		600
13+00		600
14+00		600
15+00		600

Station	Cut Volume ft^3	Fill Volume ft^3
0+00 to 1+00	− 140,000	
1+00 to 2+00	− 140,000	
2+00 to 3+00	− 150,000	
3+00 to 4+00	− 160,000	
4+00 to 5+00	− 140,000	
5+00 to 6+00	−100,000	
6+00 to 7+00	− 70,000	
7+00 to 8+00	− 40,000	
8+00 to 9+00	− 10,000	0
9+00 to 10+00	0	20,000
10+00 to 11+00		45,000
11+00 to 12+00		55.000
12+00 to 13+00		60,000
13+00 to 14+00		60,000
14+00 to 15+00		60,000
Total	− 950,000	300,000

$V_{L,waste} = -1,092,500 + 360,000$

$= -732,500 \, ft^3$

$No. of \, trips = \dfrac{732,500 \, ft^3}{11 \, yd^3 \times 27 \frac{ft^3}{yd^3}}$

$= 2,466 \, trip$

Correct Answer is (A)

SOLUTION 1.9

Company revenue:

Contract fees (company revenue)

$= \$100,000,000 \times 2.5\%$

$= \$2,500,000$

Project cost and Gross Margin (GM):

Direct labor (salaries)

$= \$80,000 \times 5 \times 1.3 \times 2.5$

$= \$1,300,000$

Truck expenses

$= \$0.25 \times 5 \times 95 \, mile \times 240 \times 2.5$

$= \$71,250$

Total project expenses

$= \$1,300,000 + \$71,250$

$= \$1,371,250$

Gross Margin (GM)

$= \$2,500,000 - \$1,371,250$

$= \$1,128,750$

Gross Margin (%)

$= \$1,128,750/\$2,500,000$

$= 45.2\%$

Company Operating Income (OI):

$OI = GM - OH - Capital \, Cost - Alloc$

Overhead cost (OH)

$$= Revenue \times OH\%$$
$$= \$2,500,000 \times 12.5\%$$
$$= \$312,500$$

Company capital costs (trucks in this case)

$$= \$55,000 \times 5$$
$$= \$275,000$$

Allocations (Alloc)

$$= Revenue \times Alloc\%$$
$$= \$2,500,000 \times 7.5\%$$
$$= \$187,500$$

$$OI = \$1,128,750 - \$312,500 - \$275,000 - \$187,500$$
$$= \$353,750$$

$$OI\,(\%) = \$353,750/\$2,500,000$$
$$= 14.2\%$$

The capital value for trucks is not usually factored into projects' Gross Margins, this is why this value was taken towards the bottom line.

Therefore, to maintain an Operating Income of:

$$OI = \$2,500,000 \times 15\% = \$375,000$$

Trucks should be salvaged at a minimum of:

$$\$375,000 - \$353,750 = \$21,250$$

$$\$21,250/5 = \$4,250 \text{ per truck}$$

Correct Answer is (C)

Solution discussion:

The proposed solution can be implemented using various avenues. However, the solution was presented in a manner consistent with the prevalent practices observed among professional organizations for their financial reporting.

In this context, company revenue (or part of it) is represented by its contract fees of $2,500,000.

The overhead cost $312,500 includes expenditures related to marketing, administration, and other non-chargeable project costs, and is customarily computed. It is reasonable to consider this cost as a percentage of the company's overall revenue, in this case 12.5%.

The same principle applies to allocations calculated as $187,500. In certain instances, particularly within large organizations, these costs may be separately itemized, especially in cases where the company operates across multiple regions, maintains various offices, and supports executive functions such as the C-suite and the President's office.

Direct labor accounts for actual staff salaries which averages at $80,000, while the 30% fringe benefits incorporate provisions for annual leave, medical leave, and other contractual benefits granted to employees.

Although trucks' capital cost was not factored into the project cost, it is noteworthy that such capital expenditures reflect a deliberate choice made by the company when alternate options, such as renting trucks, could be available. However, the assumption to include trucks capital cost within the capital expenditure framework has been made as part of the solution. Consequently, such costs do not factor into the project margin calculation.

In other instances, the company would (internally) rent out those trucks (after procuring them) to the project and reduces its project Gross Margin. None of these methods would change the final outcome of the solution.

Truck expenses (such as mileage and operation and maintenance costs = $71,250 over 2.5 *years*) are classified as project-related expenditures, and therefore reduces project's Gross Margin or its profitability.

Operating Income OI in this case $353,750 – which, is also termed as EBITDA (Earnings Before Interest, Taxes, Depreciation, and Amortization) – is derived by subtracting the overhead costs $312,500, allocations $187,500, and any other associated capital investments (exemplified by trucks in this case) $275,000, from the project's Gross Margin $1,128,750.

Considering the scenario at hand, where the company did not achieve the target Operating Income from this project $375,000, selling those trucks and realizing a minimum final sales value of $4,250 per truck could be deemed a viable option, particularly if there are no alternative contracts that necessitate the use of the trucks and given those trucks are at the end of their operating life.

SOLUTION 1.10
Project expenses include the cost of material along with the cost of borrowing, in which case you will be the one borrowing from the bank on behalf of the contractor to supply them with the material.

$$Cost\ of\ borrowing = 6\% \times \$100{,}000$$
$$= \$6{,}000$$

$$Cost\ of\ material = \$100{,}000$$

$$Total\ expenses = \$100{,}000 + \$6{,}000$$
$$= \$106{,}000$$

Overall sales inclusive of profit on all expenses
$$= \$106{,}000 \times 1.2$$
$$= \$127{,}200$$

$$Monthly\ payment = \frac{127{,}200}{12} = \$10{,}600$$

Correct Answer is (A)

SOLUTION 1.11
Chapter 7 Economic Appraisal of the *HSM Manual 1st edition* along with Chapter 1 of the *NCEES Handbook version 2.0* Section 1.7.10 Interest Tables are referred to in order to provide a solution for this question.

Start with calculating the Present Value (PV) for benefits of this project as follows:

$$PV_{benefit} = PV_{before} - PV_{after}$$

Value of annual accidents <u>*before*</u> *widening:*
$$A_{before} = 4 \times \$304{,}400 + 25 \times \$111{,}200 +$$
$$15 \times \$62{,}700 + 9 \times \$10{,}100$$
$$= \$5{,}029{,}000$$

$$PV_{before} = (P/A, 5\%, 10) \times A_{before}$$
$$= 7.7217 \times \$5{,}029{,}000$$
$$= \$38{,}832{,}429$$

<u>Value of annual accidents after widening:</u>
$$A_{after} = 2 \times \$304{,}400 + 1 \times \$111{,}200 +$$
$$4 \times \$62{,}700 + 2 \times \$10{,}100$$
$$= \$991{,}000$$

$PV_{after} = (P/A, 5\%, 10) \times A_{after}$

$= 7.7217 \times \$\,991{,}000$

$= \$\,7{,}652{,}205$

$PV_{benefit} = \$\,38{,}832{,}429 - \$\,7{,}652{,}205$

$= \$\,31{,}180{,}224$

Cost Benefit ratio calculation:

$B/C = \dfrac{PV_{benefit}}{PV_{cost}} = \dfrac{\$\,31{,}180{,}224}{\$\,15{,}000{,}000} = 2.08$

A B/C of 2.08 suggests that for every dollar invested on this road widening improvement, you would receive approximately $\$\,2.08$ in benefits over the 10 *years* study period.

Typically, a B/C greater than $'1.0'$ is considered economically viable, as it indicates that the benefits outweigh the costs.

Correct Answer is (B)

SOLUTION 1.12
The present net worth is the capitalized cost (P) for a constant annual cost over an infinite period. See Capitalized costs equation in Section 1.7.7 of the *NCEES Handbook version 2.0*.

$P = \$1{,}000{,}000 + \dfrac{75{,}000}{4\%} = \$2{,}875{,}000$

Correct Answer is (A)

SOLUTION 1.13
There are few methods that can be used to solve this question, below is one.

Start with calculating the surface area for one wall combinations as follows:

$A = 2 \times \left[\left(32 - \dfrac{14}{12}\right) \times \dfrac{14}{12}\right] + 30 \times \dfrac{14}{12}$

$= 106.9\ ft^2$

Rate of concrete placement in ft^3/hr:

$\alpha = 70\ yard^3/hr \times \dfrac{27\ ft^3}{yard^3}$

$= 1{,}890\ ft^3/hr$

In *four* hours, the volume of pumped concrete should be as follows:

$V = 1{,}890\ ft^3/hr \times 4\ hrs$

$= 7{,}560\ ft^3$

Height of concrete poured in *four* hours:

$h = \dfrac{7{,}560\ ft^3}{106.9\ ft^2} = 70.7\ ft$

Divide this by each wall combination height to determine how many walls could be poured in full:

$No.\ of\ walls = \dfrac{70.7\ ft}{11\ ft} = 6.43\ wall$

This answer represents *six* walls fully poured and 43% of the seventh wall being poured which equals to $43\% \times 11 = 4.73\ ft$

Correct Answer is (C)

SOLUTION 1.14
The effective annual interest rate is the adjusted annual rate after compounding over a given period. See Nonannual Compounding and effective interest rate equation in Section 1.7.2 of the *NCEES Handbook version 2.0*.

$i_e = \left(1 + \dfrac{r}{m}\right)^m - 1 = \left(1 + \dfrac{7\%}{4}\right)^4 - 1 = 7.19\%$

Correct Answer is (A)

SOLUTION 1.15

The depreciation rate is calculated using annual depreciation (straight line method can be used) divided by the depreciable cost.

Using *NCEES Handbook version 2.0* Section 1.7.5, straight line depreciation is computed as follows:

$$D_j = \frac{C - S_n}{n}$$

$$= \frac{\$125{,}000 - \$10{,}000}{10}$$

$$= \$11{,}500$$

$$\text{Rate of depreciation} = \frac{\$11{,}500}{\$125{,}000 - \$10{,}000}$$

$$= 10\%$$

Correct Answer is (B)

TRAFFIC ENGINEERING
Capacity Analysis, Transportation Planning, and Safety Analysis

Knowledge Areas Covered

SN	Knowledge Area
2	**Traffic Engineering (Capacity Analysis, Transportation Planning, and Safety Analysis)** A. Uninterrupted flow (e.g., level of service [LOS], capacity) B. Street segment interrupted flow (e.g., level of service [LOS], running time, travel speed) C. Intersection capacity (e.g., at grade, signalized, roundabout, interchange) D. Traffic analysis (e.g., volume studies, peak hour factor, speed studies, modal split, trip generation, traffic impact studies) E. Traffic safety analysis (e.g., conflict analysis, crash rates, collision diagrams) F. Nonmotorized facilities analysis (e.g., pedestrian, bicycle) G. Traffic forecasts and monitoring H. Highway safety analysis (e.g., crash modification factors, Highway Safety Manual)
7	**Traffic Signals** A. Traffic signal timing (e.g., clearance intervals, phasing, pedestrian crossing timing, railroad preemption) B. Traffic signal warrants C. Traffic signal design
8	**Traffic Control Design** A. Permanent signs and pavement markings B. Temporary traffic control

PART II
Traffic Engineering

PROBLEM 2.1 *Vehicles Average Spacing*

The average spacing between vehicles in a highway that has a flow rate of 1,500 veh/h and an average travel speed of 65 mph is most nearly:

(A) 105 ft/veh

(B) 23 ft/veh

(C) 40 ft/veh

(D) 230 ft/veh

PROBLEM 2.2 *Freeway Lane Width*

Determine the absolute minimum lane width to be used for a three-lane, one-way basic freeway segment that is expected to receive 80,000 vehicles per day as an Average Annual Daily Traffic with a Level of Service LOS C.

Busses are expected to form 6% of this traffic. The proportion of the traffic moving in the peak direction of travel during peak hour is 60%, and the proportion of AADT that occurs during peak hour is 8.2%.

Speed limit for this freeway is 65 mph, no ramps provided, and no obstructions of whatsoever. The Peak Hour Factor $PHF = 0.90$, and the roadway is level.

(A) 9 ft

(B) 10 ft

(C) 11 ft

(D) 12 ft

(⁂) PROBLEM 2.3 *Changing Freeway Density*

Determine the change in a four lane – two way each direction – freeway density due to an increase in its traffic volume from 2,000 veh/h to 3,000 veh/h.

The following data can be used:

- $FFS = 65\ mph$.
- Traffic is composed of 4% trucks and is located on a rolling terrain.
- Peak Hour Factor $PHF = 0.90$.
- The freeway experiences intermittent periods of moderate rainfall.

Given the above information, the change in the freeway density in *pcpmpl* is most nearly:

(A) 12

(B) 10

(C) 5

(D) 9

PROBLEM 2.4 *Multilane Highway LOS*

The below is a plan view for a 1.25 $mile$ long undivided multilane highway with a 3% upgrade at the South-North direction. Traffic composition is 7% trucks with a truck mix of 70% SUTs and 30% TTs.

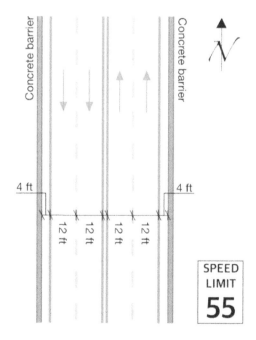

There are *five* access points per *mile* for both directions. The Peak Hour Factor $PHF = 0.95$

and peak volume in the South-North direction is 3,000 veh/hr.

The LOS which the South-North direction is expected to operate at is:

(A) LOS A

(B) LOS B

(C) LOS C

(D) LOS D

(✻) PROBLEM 2.5 *Managed Lane Volume*
The below two freeway designs are studied for the same freeway segment. The peak hourly volume at the HOV lane for Design A is estimated as 2,000 veh/hr. Free-Flow Speed of $FFS = 60\ mph$ applies to the entire freeway for the two designs.

Peak Hour Factor $PHF = 0.9$, no trucks are allowed on the HOV lane, $f_{HV} = 0.85$ for the General-Purpose lanes and their density is 27 $pcpmpl$.

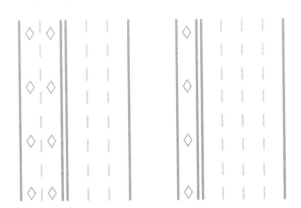

Design A   Design B

The reduction in the peak volume for the HOV lanes for Design B due to the change in lanes' arrangements from Design A to B is most nearly:

(A) 1,870 veh/hr

(B) 1,230 veh/hr

(C) 844 veh/hr

(D) 1,950 veh/hr

(✻) PROBLEM 2.6 *LOS for Two-Lane Highway*
The below 1 $mile$ two-lane highway, with a rolling terrain, passes through and serves a small town with an estimated traffic composition of 7% trucks and 9% recreational vehicles.

This highway segment has *six* access points to its east side and *five* to its west.

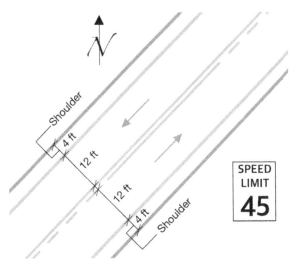

The most representative snippet of demand volumes measured in 15 min increments is summarized as follows:

Time	SW-NE direction $Veh/15\ m$	NE-SW direction $Veh/15\ m$
4:00 PM	205	195
4:15 PM	195	180
4:30 PM	196	165
4:45 PM	165	200
5:00 PM	207	197
5:15 PM	210	134
5:30 PM	195	165

Given the no-passing zones make 40% in the two directions. The LOS the SW-NE direction is expected to operate at is most nearly:

(A) LOS A

(B) LOS B

(C) LOS C

(D) LOS D

PROBLEM 2.7 *Bike Lane LOS*

The below two-lane highway's shoulders are proposed to be repurposed into bicycle lanes. The hourly directional volume is 250 veh/hr, $PHF = 0.9$, pavement rating is $'4'$ and percentage of heavy traffic is 3%.

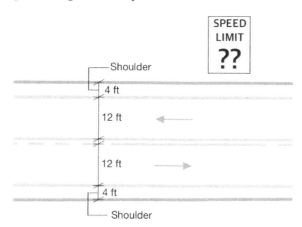

As part of this repurposing, no parking will be allowed on this highway.

The posted speed limit that can be used on this highway to maintain a bicycle LOS C should be:

(A) 40 mph

(B) 50 mph

(C) 60 mph

(D) 70 mph

PROBLEM 2.8 *Average Pedestrian Space*

Sidewalks in an urban area are being investigated for performance. An urban 1,000 ft street segment is being analyzed and has the following attributes:

- 12 ft wide sidewalk.
- Pedestrian flow rate 4,000 p/hr.
- There are no objects such as trees or anything else inside or outside the sidewalk.
- Shops alongside the sidewalk has nearly 300 ft of its length as window displays and the rest consists of building facades.

Assuming less than 20% of pedestrians are elderly, the expected average pedestrian space is most nearly:

(A) 22 ft^2/p

(B) 31 ft^2/p

(C) 35 ft^2/p

(D) 41 ft^2/p

(⁂) PROBLEM 2.9 *Crossing Difficulty Factor*

Which of the following contributes to increasing the roadway crossing difficulty factor in urban streets:

I. Increase in the number of through lanes in the subject direction.
II. Increase in the distance from crossing point to the nearest signal-controlled crossing.
III. Increase in vehicles running speed.
IV. Increase in midsegment demand flow rate in the direction nearest to the sidewalk.

(A) I + II

(B) I + III + IV

(C) II + III + IV

(D) I + II + III + IV

PROBLEM 2.10 *Bicycle Meeting Events*

The bicycle meeting events with pedestrians in an off-street, shared use, 12 ft wide path with the following data:

- Bicycles volume both directions 75 $bicycles/hr$, speed = 16.0 ft/sec
- Pedestrians' volume = 300 ped/hr
- Pedestrians mean speed = 4.3 ft/sec
- Peak Hour Factor $PHF = 0.85$

(A) 112 *events/hr*

(B) 168 *events/hr*

(C) 56 *events/hr*

(D) 4 *events/hr*

PROBLEM 2.11 *Off-Street Walkway*

The volume-to-capacity ratio and Level of Service for an off-street walkway with an effective width of 5 ft, volume of pedestrian 1,200 p/h, platooning at a speed of 4.0 ft/s, and $PHF = 0.83$, is most nearly:

(A) 0.2, *LOS C*

(B) 67, *LOS C*

(C) 1445, *LOS F*

(D) 0.3, *LOS C*

(✽) PROBLEM 2.12 *Transit Travel Speed*

The travel speed for a transit in an urban area located on a 2,000 ft segment, 40 $seconds$ segment running time and 25 $seconds$ dwell time, with one stop (*) on the segment is most nearly:

(A) 37 *mph*

(B) 34 *mph*

(C) 17 *mph*

(D) 21 *mph*

(*) The stop is not near a boundary intersection. There is no intersection at this segment and the single stop is an on-line stop (i.e., there is no reentry to a traffic stream).

PROBLEM 2.13 *Rapid Transit Max Speed*

The maximum speed for a rapid transit motorized vehicle operating in a 1.5 $mile$ long segment that should take the transit 1.5 $minutes$ to pass through (start and stop), with an acceleration of 5 ft/sec^2 and a deceleration of 4.5 ft/sec^2, is most nearly:

(A) 60 *mph*

(B) 84 *mph*

(C) 74 *mph*

(D) 96 *mph*

PROBLEM 2.14 *Lane Capacity*

The lane capacity for a single exclusive one-way through lane at a signalized intersection with the following attributes is most nearly:

o Average lane width is 10 ft.
o On-street parking adjacent to the subject lane allowed with an estimate of 10 parking maneuvers per hours.
o Zero grade for this lane.
o Heavy traffic estimated at 7%.
o Location is in a metropolitan business district with population > 300,000.
o Cycle length 100 $Seconds$, effective green time 20 $Seconds$.

(A) 276 *veh/h*

(B) 380 *veh/h*

(C) 1,534 *veh/h*

(D) 1,382 *veh/h*

(✽) PROBLEM 2.15 *Adjusted Saturation Rate for a Permitted Right Turn*

The adjusted saturation flow rate for the following shared lane with a West to East permitted right turn at a signalized intersection that has the following attributes is most nearly:

o Average width of lane is 9 ft.
o Negative grade & approach of −2%.
o Heavy traffic estimated at 10%.
o Location is in a metropolitan business district with population > 250,000.
o Cycle length 120 $seconds$, effective green time 20 $seconds$, pedestrian service time 10 $seconds$.

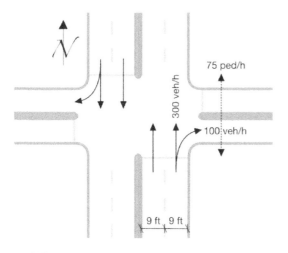

(A) 1,462 $veh/h/ln$

(B) 1,394 $veh/h/ln$

(C) 1,576 $veh/h/ln$

(D) 1,340 $veh/h/ln$

PROBLEM 2.16 *Adjusted Saturation Rate for a Permitted Left Turn*

The adjusted saturation flow rate for the East-West permitted left-turn at a signalized intersection that has the following attributes is most nearly:

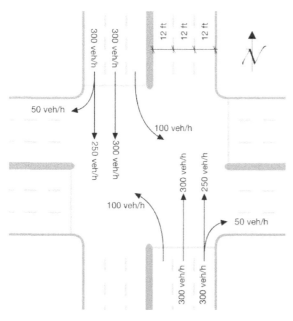

- Average width of lane is 12 ft.
- Grade of 0%
- Heavy traffic estimated at 10%
- Location is in a metropolitan business district with population > 250,000.
- Assume no conflict with pedestrians for left-turning vehicles.

(A) 871 veh/h

(B) 971 veh/h

(C) 518 veh/h

(D) 677 veh/h

PROBLEM 2.17 *Queuing Diagram*

Traffic is arriving at an intersection at a rate of *five* vehicles per minute. An accident that occurred on the intersection reduced the rate of departure of traffic to *two* vehicles per minute. The accident cleared out after 20 *minutes* and officers came in and started facilitating departing traffic to dissipate at a rate of *eleven* vehicles per minute.

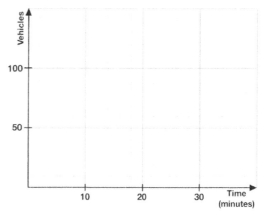

Using a queuing diagram, the maximum queue which would accumulate due to this accident, and the time it takes to clear out this queue and get back to the original condition is most nearly (*):

(A) 150 veh, 30 $minutes$

(B) 70 veh, 12 $minutes$

(C) 60 veh, 10 $minutes$

(D) 12 veh, 12 $minutes$

(*) Check question extras in the solution section for more information.

PROBLEM 2.18 *Weaving Segment Capacity*

The following weaving segment has no heavy vehicles or trucks allowed. Vehicles flow rates during the 15 *minutes* peak is tabulated below. Free Flow Speed for the freeway is 60 *mph*.

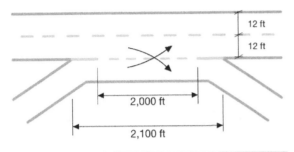

Direction of flow	Flow rate v_i in *pcph*
Freeway to Freeway	700
Freeway to Ramp	150
Ramp to Freeway	850
Ramp to Ramp	50

Based on the above information, the weaving segment capacity is most nearly:

(A) 1,405 *pcphpl*

(B) 1,790 *pcphpl*

(C) 4,215 *pcphpl*

(D) 7,164 *pcphpl*

(✻) PROBLEM 2.19 *Managed Lane Weaving Segment Intensity Factor*

The following Managed Lane weaves with the General Purpose two lanes with the following weaving segment characteristics:

- No heavy vehicles or trucks allowed.
- Vehicles flow rates during the 15 *minutes* peak is tabulated below.
- Free Flow Speed for the freeway is $FFS = 65$ *mph*.
- There is one interchange 4 *miles* before this segment and another one 4 *miles* after it.

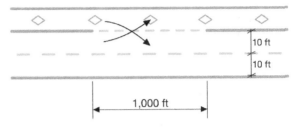

Direction of flow	Flow rate v_i in *pcph*
Freeway to Freeway	1,700
Freeway to ML	750
ML to Freeway	500
ML to ML	200

Based on the above information, the weaving intensity factor is most nearly (*):

(A) 0.22

(B) 0.34

(C) 0.51

(D) 0.83

(*) Check question extras in the solution section for more information.

PROBLEM 2.20 *Crosswalk Occupancy Time*

Crossing A width is 14 *ft* and has pedestrians flowing as shown in the below diagram.

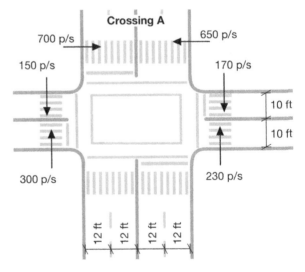

Given the following information:
- Cycle length 110 *seconds*

- Walking phases for all crossings is set at 10 *seconds*
- Walking speed 4 *ft/sec*

The crossing occupancy time for the major Crossing A is most nearly:

(A) 772 *p.s*

(B) 353 *p.s*

(C) 675 *p.s*

(D) 1,072 *p.s*

PROBLEM 2.21 *Merging Segment Density*
The below are two merging lanes in a *six*-lane freeway.

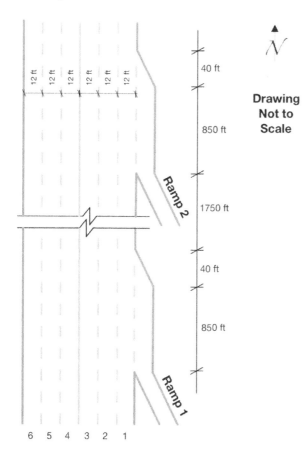

Considering the following flow rates:

Segment	Flow rate
	pcph
Freeway	4,000
Ramp 1	350
Ramp 2	350

Density in ramp 2 influence area in *pcpmpl* is expected to be:

(A) 10

(B) 23

(C) 35

(D) 41

PROBLEM 2.22 *Roundabout bypass Lane*
The below is a single lane roundabout with single lanes for entrances and exits along with a yielding bypass on its northbound direction.

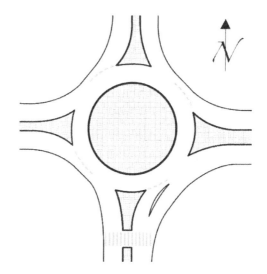

Demand volumes in *veh/h* are as follows:

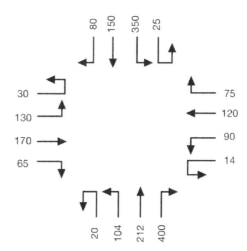

Considering the following information:

- Heavy traffic not permitted.
- Peak Hour Factor $PHF = 1.0$
- Pedestrians on the south leg 75 p/h and negligible everywhere else.

The volume to capacity ratio for the bypass lane is most nearly:

(A) 0.51

(B) 0.94

(C) 0.76

(D) 0.63

PROBLEM 2.23 *Two-Way Stop-Controlled Intersection*

The below is a Two-Way Stop-Controlled Intersection with the main road at the East-West direction and the minor road, which is supplied with stop signs, at the South-North direction.

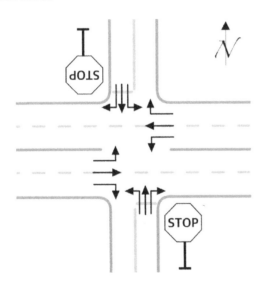

Demand volumes in veh/h are as below, the Peak Hour Factor $PHF = 0.9$, negligible pedestrians' traffic assumed, no grade, and no heavy traffic allowed:

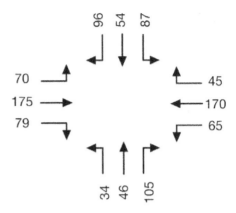

The potential capacity for the left turning lane heading from West and turning towards a Northbound trip (the one with 70 veh/h) is most nearly:

(A) 1,615 veh/h

(B) 1,400 veh/h

(C) 5,515 veh/h

(D) 410 veh/h

PROBLEM 2.24 *Dilemma Zone Elimination*

With a 40 ft wide intersection, average vehicles length of 14 ft and an assumed driver perception time of 1.5 sec, along with a yellow interval of 4 sec with acceleration and deceleration rates of 10 ft/sec, the range of speed that can avoid a dilemma zone is most nearly:

(A) 8 mph to 26 mph

(B) 8 mph to 38 mph

(C) $\leq$ 26 mph

(D) $\leq$ 38 mph

PROBLEM 2.25 *Diverging Diamond Interchange DDI*

Which of the following statements is/are true for Diverging Diamond Interchanges DDIs:

I. DDIs have freeway entry and exit ramps separated at the street level creating four intersections.
II. Large efficiency gains for interchanges with heavy left-turn demand.
III. Some unique considerations need to take place for lane utilization and saturation flow rate adjustments for crossovers and U-turn movements.
IV. DDIs allow free flowing left-turn movements onto the crossing freeway.

(A) I + II + III + IV
(B) III + IV
(C) II + III + IV
(D) II + IV

PROBLEM 2.26 *Traffic Signal Warrant*

The below is a $4\,ft\,8.5\,in$ wide, highway-LRT grade, crossing a minor two-lane street at an angle as shown.

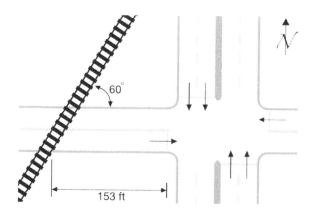

The volume of traffic on each approach of the intersection along with other traffic data regarding rail is summarized below:

Facility	Major Rd. N-S	Major Rd. S-N	Minor Rd. E-W	Minor Rd. W-E	Rail
Width ft	48	48	24	24	4.7
Speed mph	45	45	45	45	25
%age of busses > 20 passengers	4	4	4	4	–
%age of Tractor-Trailer Trucks	2	3	15	16	–
Hour	Volume Veh/h				
8AM – 9AM	193	181	93	98	2
9AM – 10AM	181	116	56	102	
10AM – 11AM	117	143	82	33	
11AM – 12AM	88	115	43	56	
12AM – 1PM	151	189	102	98	
1PM – 2PM	190	83	36	15	2
2PM – 3PM	93	93	56	33	
3PM – 4PM	87	93	97	56	
4PM – 5PM	203	113	102	99	2

Based on this information, which statement is true regarding traffic control for rail and intersection:

A) A traffic signal is warranted for this intersection along with automatic gates for the rail.

B) A traffic signal is not warranted for this intersection. Automatic gates for the rail maybe required.

C) A traffic signal is not warranted for this intersection. Automatic gates for the rail must be installed.

D) A traffic signal is warranted for this intersection. Automatic gates maybe required.

PROBLEM 2.27 *Interior Lane Closure*

The below is a low volume urban four-lane street with one internal lane closed to account for work to be done in the indicated hatched area:

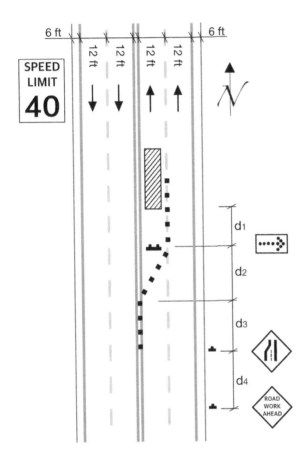

Based on the information given above, distances $d1$, $d2$, $d3$ and $d4$ respectively in ft should be as follows:

(A) 305, 320, 100, 100

(B) 0, 100, 100, 50

(C) 305, 160, 100, 100

(D) 305, 107, 100, 100

PROBLEM 2.28 *Chevron Sign Alignment*

Chevron signs are to be placed on an exit ramp with a horizontal curve radius of 150 ft. The speed limit of the ramp is 15 mph.

The recommended chevron sign spacing in this case is:

(A) Chevrons are optional at this speed

(B) 40 ft

(C) 80 ft

(D) 30 ft

PROBLEM 2.29 *Signal Offset and Cycle Length*

The cycle length for an ideal two-way progression for intersection No. 2 below that generates a moving platoon at an offset of 10 *seconds* is most nearly:

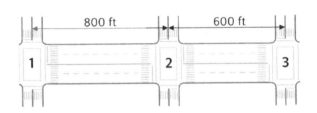

(A) 47 *sec*

(B) 14 *sec*

(C) 10 *sec*

(D) 20 *sec*

PROBLEM 2.30 *Coordinated Signal System*

The following statement(s) are true regarding progressive signals design for a coordinated signal system:

I. For traffic control signals when coordinated, a sign should be used to identify the coordinated speed.
II. Signs used to identify a coordinated signals system can have a changeable message for speed and time if those were not consistent during the day/week.
III. They are installed predominantly for one way direction streets to provide the necessary degree of vehicular platooning.
IV. They are best applied when the resultant spacing between signals is less than 1,000 ft.

(A) I + III + IV

(B) I + II + IV

(C) II

(D) II + IV

PROBLEM 2.31 *Uniform Delay for a Signalized Intersection Through Lane*

The uniform delay for a signalized intersection through lane that has a cycle length of 100 sec and an effective green time of 20 sec, saturation of 1,750 veh/h and a flow rate of 300 veh/h with an arrival type 4 and a progression that is described as good progression, is most nearly:

(A) 39 sec

(B) 27 sec

(C) 23 sec

(D) 33 sec

PROBLEM 2.32 *Lane Reduction Signs and Road Marking*

The following plan view represents a lane reduction for what used to be a heavy two-lanes traffic moving at a speed of 60 mph northbound reduced to one-lane of 40 mph:

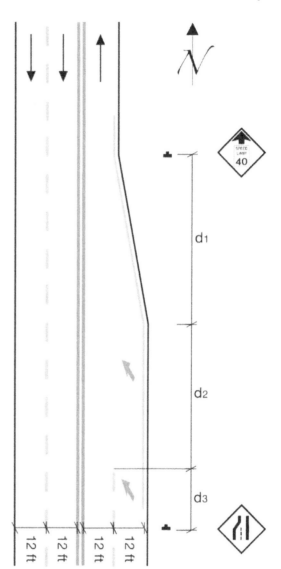

Distances $d1$, $d2$ and $d3$ in ft should be as follows respectively:

(A) 720, 825 & 275

(B) 720, 1100 & 275

(C) 320, 500 & 170

(D) 320, 825 & 275

PROBLEM 2.33 Crash Rate

The Annual Average Daily Traffic AADT for four intersections 1, 2, 3 and 4 is presented in the following table along with their total observed crashes during a study period of five years.

Intersection	Major AADT	Minor AADT	Crashes per five yrs.
1	17,800	12,100	97
2	19,500	5,100	46
3	22,000	6,300	30
4	13,000	2,500	21

The following ranking is the best representation for those intersections from their highest crash rate to their lowest crash rate:

(A) 1, 2, 4, 3

(B) 1, 2, 3, 4

(C) 3, 4, 2, 1

(D) 4, 3, 2, 1

PROBLEM 2.34 Equivalent Property Damage Only (EPDO) Score

The table below represents the societal crash cost assumptions as updated by the FHWA for the year 2016. QALY represents Quality Adjusted Years – i.e., intangible pain and suffering cost.

Crash severity	Economic Crash Unit Cost	QALY Crash Unit Cost	Crash Unit Cost
Fatal (K)	$1,688,100	$4,052,000	$5,740,100
Disability (A)	$151,000	$153,400	$304,400
Evident (B)	$56,800	$54,400	$111,200
Possible (C)	$38,500	$24,200	$62,700
PDO (O)	$8,700	$1,400	$10,100

Intersection "x" experienced the following crash types during the study period:

Crash severity	Crashes
Fatal (K)	3
Disability (A)	4
Evident (B)	13
Possible (C)	33
PDO (O)	47

Based on this information, the EPDO Score for this intersection is most nearly:

(A) 920

(B) 9,828

(C) 3.6

(D) 2,212

PROBLEM 2.35 Roadway Sideslope Modification

An existing sideslope of 1V:2H was reduced to 1V:7H in a bid to improve traffic safety for a segment that was experiencing a total of 55 *crashes per year*.

The expected reduction in crashes per year due to this countermeasure is most nearly:

(A) 8

(B) 46

(C) 40

(D) 15

PROBLEM 2.36 *Countermeasures Proposed*

A road segment has been experiencing numerous vehicles rollover over the course of several years.

The following countermeasure(s) could be best implemented to tackle the issue at hand:

   I. Increase shoulder width
   II. Assess and address any drainage issues which could result in slippery surface
   III. Assess and improve road markings
   IV. Assess and improve signage

(A) I + II + III + IV

(B) I

(C) III + IV

(D) III

PROBLEM 2.37 *Trip Generation Model*

The following table shows the number of trips per household per day for different combinations of household size and vehicle ownership in a certain community:

Household size	Vehicles ownership		
	0	1	2 +
1	3	4	6
2	4.5	4.5	8
3	7	8.5	9
4	8	9.1	11
5 +	10.4	13	14.5

The forecasted number of households in the study period is as follows:

Household size	Vehicles ownership		
	0	1	2 +
1	100	150	200
2	150	200	250
3	200	250	300
4	250	300	350
5 +	300	350	400

Based on this, the forecasted number of daily trips from this community is most nearly:

(A) 33,950 *trips per day*

(B) 13,750 *trips per day*

(C) 3,750 *trips per day*

(D) 26,455 *trips per day*

PROBLEM 2.38 *Roadway Density*

The Density of a highway measured in $veh/mile$ that has a flow rate of $1,500\ veh/hr$ and an average travel speed of $65\ mph$ is most nearly:

(A) 105

(B) 34

(C) 40

(D) 23

PROBLEM 2.39 *Peak Hour Factor*

In the below traffic flow count, the Peak Hour Factor PHF for the SW-NE direction is 5% higher than the NE-SW direction.

Time	SW-NE direction $Veh/15\ m$	NE-SW direction $Veh/15\ m$
4:00 PM	205	195
4:15 PM	195	180
4:30 PM	196	165
4:45 PM	165	X
5:00 PM	207	197
5:15 PM	210	134
5:30 PM	195	125

Given the information above, the estimated 15 *min* traffic flow in the NE-SW direction at 4:45 PM identified as 'X' is most nearly:

(A) 180 *Veh*

(B) 215 *Veh*

(C) 197 *Veh*

(D) 236 *Veh*

PROBLEM 2.40 Transit Average Speed

The average speed for a rapid transit motorized vehicle which accelerates from a station at a rate of $5\ ft/sec^2$ to reach a maximum speed of $90\ mph$, stays on this speed for $1.5\ minutes$ before it decelerates at a rate of $4.5\ ft/sec^2$ into its next stop is most nearly:

(A) 60 mph
(B) 56 mph
(C) 73 mph
(D) 51 mph

PROBLEM 2.41 Skid Marks Distance

Two cars travelling at a level road in opposite directions to each other, the first car is travelling at a speed of $90\ mph$ and the second car at a speed of $60\ mph$. Due to reduced vision they both came into this realization when they were only $1{,}340\ ft$ apart. With a perception time of $2\ seconds$, they both applied brakes and came into a stop leaving a gap of $35\ ft$ between them.

The distance traveled by the first car and the second car respectively after applying brakes is most nearly:

(A) 897 ft, 468 ft
(B) 599 ft, 266 ft
(C) 519 ft, 346 ft
(D) 540 ft, 325 ft

PROBLEM 2.42 85th Percentile Speed

The below table has 21 speed readings measured in *mph* for vehicles moving passed a monitoring point:

55	59	46	45	69
45	32	12	21	59
72	25	56	45	73

The 85th percentile speed for the above data set is:

(A) 69 mph
(B) 48 mph
(C) 25 mph
(D) 64 mph

PROBLEM 2.43 Yellow Interval

The yellow change interval is being designed for a $48\ ft$ wide level approach. Considering the design speed in that location is $45\ mph$, the best timing for the yellow phase is:

(A) 3.3 seconds
(B) 4.3 seconds
(C) 5.3 seconds
(D) 6.3 seconds

PROBLEM 2.44 Toll Sign Board

A toll auxiliary sign is to be placed above a route sign to inform that a particular conventional road has a toll gate ahead. The minimum size in *inches* for this signboard should be:

(A) 24 × 12
(B) 36 × 18
(C) 96 × 66
(D) 114 × 66

PROBLEM 2.45 School Crossing Signal

The following is factored into the decision of installing a school crossing signal – select all that applies:

(A) The need for a traffic study.
(B) The minimum number of all kids during peak hour.
(C) Frequency of gaps in the traffic system across the major street next to school.
(D) The distance to the nearest traffic signal.

SOLUTION 2.1

Referring to Chapter 4 of the *HCM Manual* 6th edition Equations 4-6 and 4-7, the following can be determined:

$$Flow\ rate\ (veh/h) = \frac{3{,}600\ s/h}{av.\ headway\ (s/veh)}$$

$$\rightarrow Av.\ headway\ (s/veh) = \frac{3{,}600\ s/h}{Flow\ rate\ (veh/h)}$$

$$= \frac{3{,}600\ s/h}{1{,}500\ veh/h}$$

$$= 2.4\ s/veh$$

$$Av.\ headway\ (s/veh) = \frac{av.\ spacing\ (ft/veh)}{av.\ travel\ speed\ (ft/s)}$$

$$2.4\ s/veh = \frac{av.\ spacing\ (ft/veh)}{65\ mph \times \frac{5{,}280\ ft/mile}{3{,}600\ h/s}}$$

$$\rightarrow Av.\ spacing\ (ft/veh)$$

$$= 2.4\ s/veh \times 65\ mph \times \frac{5{,}280\ ft/mile}{3{,}600\ h/s}$$

$$= 228.8\ ft/veh$$

A quicker method can be used by taking advantage of units as follows:

$$Av.\ spacing\ (ft/veh) = \frac{65\ mph}{1{,}500\ veh/h} \times 5{,}280\ \frac{ft}{mile}$$

$$= 228.8\ ft/veh$$

Correct Answer is (D)

SOLUTION 2.2

The width of the requested lane can be determined using Equation 12-2 along with Exhibit 12-20 from the *HCM Manual* 6th edition as follows:

$$FFS = BFFS - f_{LW} - f_{RLC} - 3.22 \times TRD^{0.84}$$

f_{RLC} is the adjustment factor for right-side lateral clearance. Given there is none in the question, this factor equals to zero. Same applies to TRD which is the total ramp density in the segment.

The Base Free-Flow Speed equals the speed limit plus $5\ mph$ as advised in Exhibit 10-7 (or page 12-28) $\rightarrow BFFS = 65 + 5 = 70\ mph$.

Provided the above, the adjustment factor for lane width f_{LW} is calculated as follows:

$$f_{LW} = BFFS - FFS$$

$$= 70\ mph - FFS$$

Use Equation 12-20 of the manual to determine Volume as follows:

$$V = DDHV = AADT \times K \times D$$

Where D, the directionality factor, is the proportion of the traffic moving in the peak direction ($= 60\%$), and K is the proportion of $AADT$ that occurs during peak hour ($= 8.2\%$).

$$V = 80{,}000 \times 0.082 \times 0.6 = 3{,}936\ vph$$

Use Equation 12-9 to determine the demand flow rate v_p as follows:

$$v_p = \frac{V}{PHF \times f_{HV} \times N}$$

f_{HV} is the heavy vehicle adjustment factor and is calculated as follows:

$$f_{HV} = \frac{1}{1 + P_T(E_T - 1)}$$

P_T is the proportion of heavy vehicles which is 6% of buses in this case, and E_T is the Passenger Car Equivalent PCE taken from Exhibit 12-25 as 2.0 for level terrains.

$$f_{HV} = \frac{1}{1 + 0.06 \times (2-1)} = 0.94$$

$$v_p = \frac{V}{PHF \times f_{HV} \times N}$$

$$= \frac{3{,}936\ vph}{0.9 \times 0.94 \times 3}$$

$$= 1{,}551\ pcphpl$$

Use Exhibit 12-37 to determine FFS. In this case, plug in the above determined flow rate v_p per lane under $LOS\ C$, and without interpolation as it is not permitted → $FFS = 60\ mph$.

$f_{LW} = 70\ mph - FFS = 10\ mph$

Using Exhibit 12-20 of the manual, the minimum lane width that can be used in this case is $10\ ft$.

Correct Answer is (B)

SOLUTION 2.3
From Exhibit 11-20 of the *HCM Manual 6th edition*, the Capacity Adjustment Factor that reflects a moderate rainfall weather condition for a $FFS = 65\ mph$ is $CAF = 0.92$, and the Speed Adjustment Factor $SAF = 0.94$.

Based on the above information and using Exhibit 12-6 of the manual, the following can be established:

$FFS_{adj} = FFS \times SAF$

$= 65\ mph \times 0.94$

$= 61.1\ mph$

The base segment capacity for $55 \leq FFS \leq 75$ and $c \leq 2,400$ is calculated as follows:

$c = 2,200 + 10 \times (FFS - 50)$

$= 2,200 + 10 \times (65 - 50)$

$= 2,350\ pcphpl$

$c_{adj} = c \times CAF$

$= 2,350 \times 0.92$

$= 2,162\ pcphpl$

Flow rate is calculated for each of the given volumes as follows:

$v_p = \dfrac{V}{PHF \times f_{HV} \times N}$

$f_{HV} = \dfrac{1}{1 + P_T(E_T - 1)}$

P_T is the proportion of heavy vehicles which is 4% of trucks in this case, and E_T is the Passenger Car Equivalent PCE taken from Exhibit 12-25 as 3.0 for rolling terrains.

$= \dfrac{1}{1+0.04\times(3-1)} = 0.93$

$v_{p,2000\ vph} = \dfrac{2,000}{0.90\times0.93\times2} = 1,195\ pcphpl$

$v_{p,3000\ vph} = \dfrac{3,000}{0.90\times0.93\times2} = 1,792\ pcphpl$

Going back to Equation 12-1 and Exhibit 12-6 to calculate speed, the breakpoint in $pcphpl$ is determined as follows:

$BP_{adj} = [1,000 + 40 \times (75 - FFS_{adj})] \times CAF^2$

$= [1,000 + 40 \times (75 - 61.1)] \times 0.92^2$

$= 1,317\ pcphpl$

Since $v_{p,2000\ vph} \leq BP$, the corresponding speed should be calculated as follows:

$S_{2000\ vph} = FFS_{adj} = 61.1\ mph$

Provided that the calculations in this solution section have shown $BP < v_p \leq c$, the corresponding speed is calculated as follows:

$S_{3000\ vph} = FFS_{adj} - \dfrac{\left(FFS_{adj} - \frac{c_{adj}}{D_c}\right)(v_p - BP)^a}{(c_{adj} - BP)^a}$

$= 61.1 - \dfrac{\left(61.1 - \frac{2,162}{45}\right)(1,792 - 1,317)^2}{(2,162 - 1,317)^2}$

$= 57\ mph$

The requested Densities are calculated using Equation 12-11 of the manual as follows:

$D = \dfrac{v_p}{S}$

$$D_{2000\,vph} = \frac{v_{p,2000vph}}{S_{2000\,vph}}$$

$$= \frac{1{,}195\ pcphpl}{61.1\ mph}$$

$$= 19.6\ pcpmpl$$

$$D_{3000\,vph} = \frac{v_{p,3000vph}}{S_{3000\,vph}}$$

$$= \frac{1{,}792\ pcphpl}{57\ mph}$$

$$= 31.4\ pcpmpl$$

$$\Delta D = D_{3000\,vph} - D_{2000\,vph}$$

$$= 31.4 - 19.6$$

$$= 11.8\ pcpmpl$$

Correct Answer is (A)

SOLUTION 2.4
In reference to the *HCM Manual 6th edition* Chapter 12, start with calculating the Free-Flow Speed using Equation 12-3 as follows:

$$FFS = BFFS - f_{LW} - f_{TLC} - f_M - f_A$$

The Base Free-Flow Speed $BFFS$ is estimated using the same manual Page 12-28 (or Exhibit 10-7) as the speed limit plus 5 mph for speed limits higher than 50 mph:

$$BFFS = 55 + 5 = 60\ mph$$

f_{LW} is the adjustment factor for lane width which equals to zero for 12 ft wide lanes.

f_{TLC} is the adjustment factor for the Total Lateral Clearance taken from Exhibit 12-22 using Equation 12-4 as follows:

$$TLC = LC_R + LC_L$$

TLC being the Total Lateral Clearance, LC_R is the right-side lateral clearance $= 4\ ft$ given there is a concrete barrier placed $< 6\ ft$ to the right-side of the lane. LC_L is the left-side lateral clearance $= 6\ ft$ for undivided highways.

$$TLC = 4\ ft + 6\ ft = 10\ ft$$

Using Exhibit 12-22, a four-lane highway with a $TLC = 10\ ft$ has a $f_{TLC} = 0.4\ mph$.

f_M is the adjustment factor for median type taken from Exhibit 12-23 as 1.6 mph.

f_A is the adjustment factor for access points density taken from Exhibit 12-24 using interpolation as 1.25 mph.

$$FFS = BFFS - f_{LW} - f_{TLC} - f_M - f_A$$

$$= 60 - 0 - 0.4 - 1.6 - 1.25$$

$$= 56.75\ mph$$

Capacity for a multilane highway is determined using Equation 12-7 or Exhibit 12-6 as follows:

$$c = 1{,}900 + 20 \times (FFS - 45)$$

$$= 1{,}900 + 20 \times (56.75 - 45)$$

$$= 2{,}135\ pcphpl$$

The breakpoint BP is also taken from Exhibit 12-6 as 1,400 $pcphpl$.

Flow rate for the peak volume in the South-North is calculated as follows:

$$v_p = \frac{V}{PHF \times f_{HV} \times N}$$

$$f_{HV} = \frac{1}{1 + P_T(E_T - 1)}$$

P_T is the proportion of heavy vehicles which is 7% of trucks in this case. E_T is the Passenger Car Equivalent PCE taken from Exhibit 12-28 with 70% SUTs and 30% TTs, with an upgrade of 3% in the South-North direction – using interpolation – as '3.18'.

$$= \frac{1}{1+0.07\times(3.18-1)} = 0.87$$

$$v_p = \frac{3{,}000}{0.95\times 0.87\times 2} = 1{,}815\ pcphpl$$

Speed is determined using Equation 12-1 of the manual. As the calculations in this solution section have shown, $BP < v_p \leq c$, the corresponding speed in this case is calculated as follows (*):

$$S = FFS_{adj} - \frac{\left(FFS_{adj} - \frac{c_{adj}}{D_c}\right)(v_p - BP)^a}{(c_{adj} - BP)^a}$$

$$= 56.75 - \frac{\left(56.75 - \frac{2{,}135}{45}\right)(1{,}815 - 1{,}400)^{1.31}}{(2{,}135 - 1{,}400)^{1.31}}$$

$$= 52.3\ mph$$

Density of the flow is calculated using Equation 12-11 of the manual as follows:

$$D = \frac{v_p}{S} = \frac{1{,}815}{52.3} = 34.7\ pcpmpl$$

Using Exhibit 12.15 of the manual, a density of 34.7 *pmpcpl* is associated with a Level of Service *LOS D*.

Correct Answer is (D)

(*) Given no adjustment factors for weather conditions or circumstances were provided in the question → $FFS = FFS_{adj}$ and $c = c_{adj}$.

SOLUTION 2.5
Chapter 12 of the *HCM Manual 6th edition* Section 4, page 12-40, Basic Managed Lane Segments is used to solve this question.

Design A is the 'Buffer 2' *HOV* segment type. Design B is the 'Buffer 1' *HOV* segment type. (Exhibit 12-9). Based upon which the corresponding Equations' 12-13 and 12-14 coefficients along with the speed equations' coefficients are taken from Exhibit 12-30.

The strategy for this solution is to find the space mean speed S_{ML} for Design A using its volume, then substitute it in Design B to find its volume, then recalculate it for confirmation.

Freeway Segment Design A:

Managed Lane (ML) – Buffer 2:

Flow rate for the peak volume in the Managed Lane is calculated as follows:

$$v_p = \frac{V}{PHF \times f_{HV} \times N}$$

$$= \frac{2{,}000}{0.9\times 1\times 2}$$

$$\cong 1{,}111\ pcphpl$$

$$BP = [BP_{75} + \lambda_{BP}\times(75 - FFS_{adj})]\times CAF^2$$

$$= [500 + 10\times(75 - 60)]\times 1.0^2$$

$$= 650\ pcphpl$$

$$c_{adj} = CAF \times [c_{75} - \lambda_c \times (75 - FFS_{adj})]$$

$$= 1.0\times[1{,}850 - 10\times(75-60)]$$

$$= 1{,}700\ pcphpl$$

$$S_1 = FFS_{adj} - A_1 \times \min(v_p, BP)$$

$$= 60 - 0\times \min(1111, 650)$$

$$= 60\ mph$$

$$A_2 = A_2^{55} + \lambda_{A2}(FFS_{adj} - 55)$$

$$= 1.5 + 0.02\times(60 - 55)$$

$$= 1.6$$

$$S_2 = \frac{\left(S_{1,BP} - \frac{c_{adj}}{K_c^{nf}}\right)}{(c_{adj} - BP)^{A2}}(v_p - BP)^{A2}$$

$$= \frac{\left(60 - \frac{1{,}700}{45}\right)}{(1{,}700 - 650)^{1.6}}(1{,}111 - 650)^{1.6}$$

$$= 6\ mph$$

$$S_{ML} = \begin{cases} S_1 & v_p \leq BP \\ S_1 - S_2 - I_c \times S_3 & BP < v_p \leq c \end{cases}$$

The formula assigned to $BP < v_p \leq c$ is used in this case. Also, in reference to Equation 12-18, $I_c = 0$ for Buffer 2 type.

$$S_{ML} = S_1 - S_2 - 0 \times S_3$$
$$= 60 - 6 - 0$$
$$= 54 \, mph$$

Freeway Segment Design B:

Managed Lane (ML) – Buffer 1:

This section will be worked on backwards starting from the space mean speed S_{ML} equation ending up with the flow rate v_p and the requested volume V.

$$BP = [BP_{75} + \lambda_{BP} \times (75 - FFS_{adj})] \times CAF^2$$
$$= [600 + 0 \times (75 - 60)] \times 1.0^2$$
$$= 600 \, pcphpl$$

$$c_{adj} = CAF \times [c_{75} - \lambda_c \times (75 - FFS_{adj})]$$
$$= 1.0 \times [1,700 - 10 \times (75 - 60)]$$
$$= 1,550 \, pcphpl$$

$$S_{ML} = \begin{cases} S_1 & v_p \leq BP \\ S_1 - S_2 - I_c \times S_3 & BP < v_p \leq c \end{cases}$$

Assume $BP < v_p \leq c$, and $I_c = 0$ as established from the General-Purpose lane for Design B and Equation 12-18 where Density $K_{GP}(= 27) < 35 \, pcpmpl$.

$S_{ML} = 54 \, mph$ (*)

$54 = S_1 - S_2$

$$S_1 = FFS_{adj} - A_1 \times \min(v_p, BP)$$
$$= 60 - 0.0033 \times \min(v_p, 600)$$
$$= 58 \, mph$$

$\rightarrow S_2 = 4 \, mph$

$$A_2 = A_2^{55} + \lambda_{A2}(FFS_{adj} - 55)$$
$$= 1.4 - 0 \times (60 - 55)$$
$$= 1.4$$

$$S_2 = \frac{\left(S_{1,BP} - \frac{c_{adj}}{K_c^{nf}}\right)}{(c_{adj} - BP)^{A2}} (v_p - BP)^{A2}$$

$$4 = \frac{\left(58 - \frac{1,550}{30}\right)}{(1,550 - 600)^{1.4}} (v_p - 600)^{1.4}$$

$$4 = 4.29 \times 10^{-4} \times (v_p - 600)^{1.4}$$

Substituting $S_2 = 4$ generates the following equation:

$$9,324 = (v_p - 600)^{1.4}$$

$\rightarrow v_p = 1,285 \, pcphpl$

Which conforms with the initial assumption of $BP < v_p \leq c \rightarrow OK$

Flow rate for the peak volume in the HOV segment is calculated as:

$$v_p = \frac{V}{PHF \times f_{HV} \times N}$$

$$1,285 = \frac{V}{0.9 \times 1 \times 1}$$

$\rightarrow V = 1,156 \, veh/hr$

$\Delta V = 2,000 - 1,156 = 844 \, veh/hr$

Correct Answer is (C)

(*) S_{ML} should be recalculated for confirmation of assumptions. The Question is longer than an exam norm already, therefore, this step could be omitted, however calculated here for completeness as follows:

$$v_p = 1{,}285\ pcphpl$$

$$BP = 600\ pcphpl$$

$$c_{adj} = 1{,}550\ pcphpl$$

$$S_1 = FFS_{adj} - A_1 \times \min(v_p, BP)$$
$$= 60 - 0.0033 \times \min(1285, 600)$$
$$= 58\ mph$$

$$A_2 = A_2^{55} + \lambda_{A2}(FFS_{adj} - 55)$$
$$= 1.4 + 0 \times (60 - 55)$$
$$= 1.4$$

$$S_2 = \frac{\left(S_{1,BP} - \frac{c_{adj}}{K_c^{nf}}\right)}{(c_{adj} - BP)^{A_2}}(v_p - BP)^{A_2}$$
$$= \frac{\left(58 - \frac{1{,}550}{30}\right)}{(1{,}550 - 600)^{1.4}}(1{,}285 - 600)^{1.4}$$
$$= 4.29 \times 10^{-4} \times (1{,}285 - 600)^{1.4}$$
$$= 4\ mph$$

$$S_{ML} = \begin{cases} S_1 & v_p \leq BP \\ S_1 - S_2 - I_c \times S_3 & BP < v_p \leq c \end{cases}$$

The formula assigned to $BP < v_p \leq c$ is used in this case. Also, in reference to Equation 12-18, $I_c = 0$ for Buffer 2 type.

$$S_{ML} = S_1 - S_2 - 0 \times S_3$$
$$= 58 - 4 - 0$$
$$= 54\ mph \rightarrow OK$$

SOLUTION 2.6

Chapter 15 of the *HCM Manual 6th edition* Exhibit 15-6 Step 1 to Step 8 is followed to solve this problem.

The description of this two-lane highway aligns with Class III highways – check page 15-4 and 15-5 for more information. Exhibit 15-6, eliminates Steps 5 and 6 for Class III highways. LOS in this case is assessed based on *Percentage of Free-Flow Speed (PFFS)*.

Step 1: Gather Input Data

Given the demand volumes are measured in 15 min increments, flow rate for the required direction (SW-NE) is calculated as follows with the use of $PHF = 1.0$:

$$v_d = \frac{4 \times V_{d,max}}{PHF} = \frac{4 \times 210}{1.0} = 840\ pcph\ (*)$$

$$v_o = \frac{4 \times V_{o,max}}{PHF} = \frac{4 \times 200}{1.0} = 800\ pcph\ (*)$$

Base Free-Flow Speed *BFFS* is estimated using Exhibit 15-5 as follows:

$$BFFS = 45 + 10 = 55\ mph$$

Step 2: Estimate Free-Flow Speed

$$FFS = BFFS - f_{LS} - f_A$$

f_{LS} is the adjustment factor for lane width (12 ft) and shoulder width (4 ft), Exhibit 15.7 → $f_{LS} = 1.3\ mph$

f_A is the adjustment factor for access point density for the two directions (i.e., '11'), interpolating Exhibit 15-8 → $f_A = 2.8\ mph$

$$FFS = 55 - 1.3 - 2.8 = 50.9\ mph$$

Step 3: Demand Adjustment for ATS

$$v_{i,ATS} = \frac{V_i}{PHF \times f_{g,ATS} \times f_{HV,ATS}}$$

$$v_{d,ATS} = \frac{840}{f_{g,ATS} \times f_{HV,ATS}}$$

$$v_{o,ATS} = \frac{800}{f_{g,ATS} \times f_{HV,ATS}}$$

$f_{g,ATS}$ is the grade adjustment factor = 0.99 for rolling terrains using Exhibit 15-9 with $v > 800$.

$f_{HV,ATS}$ is the adjustment factor for heavy vehicles and is estimated as follows:

$$f_{HV,ATS} = \frac{1}{1 + P_T(E_T - 1) + P_R(E_R - 1)}$$

Using Exhibit 15-11 → $E_T = 1.3$ and $E_R = 1.1$

$$= \frac{1}{1 + 0.07(1.3 - 1) + 0.09 \times (1.1 - 1)}$$

$$= 0.97$$

$$\rightarrow v_{d,ATS} = \frac{840}{0.99 \times 0.97} \cong 875 \; pcph$$

$$v_{o,ATS} = \frac{800}{0.99 \times 0.97} = 833 \; pcph$$

Step 4: Estimate ATS

$ATS_d = FFS - 0.00776(v_{d,ATS} + v_{o,ATS}) - f_{np,ATS}$

$f_{np,ATS}$ is the adjustment for the percentage of no-passing zones in the analysis direction, given as 40%. Using Exhibit 15-15 with $v_{o,ATS} = 833$ & $FFS = 50.9$ → $f_{np,ATS} = 0.57$

$ATS_d = 50.9 - 0.00776(875 + 833) - 0.57$

$= 37.1 \; mph$

Step 5: Adjustments for PTSF – Skipped.

Step 5: Estimated PTSF – Skipped.

Step 7: Estimate PFFS

$$PFFS = \frac{ATS_d}{FFS} = \frac{37.1}{50.9} = 0.73$$

Step 8: Determine LOS

Using Exhibit 15-3, with Class III two-lane highway, and a $PFFS > 66.7 - 75.0$, the Level of Service this highway is expected to operate at is LOS D.

Correct Answer is (D)

(*) 'd' represents the analysis direction and 'o' represents the opposite direction.

SOLUTION 2.7

Chapter 15 of the *HCM Manual 6th edition* Exhibit 15-37 Step 1 to Step 5 are referred to in order to solve this problem. Those steps are not followed in order due to the nature of the request.

The directional demand flow rate in the outside lane v_{OL} is calculated as follows given that $N = 1$ for a two-lane highway:

$$v_{OL} = \frac{V}{PHF \times N} = \frac{250}{0.9 \times 1} \cong 278 \; veh/hr$$

The effective width of the shoulder is calculated from Equations 15-26 and 28 as follows:

$W_e = W_v + W_s - 2 \times [\%OHP(2 \; ft + W_s)]$

$W_v = W_{OL} + W_s = 12 \; ft + 4 \; ft = 16 \; ft$

And given that $\%OHP = 0$:

$\rightarrow W_e = 16 \; ft + 4 \; ft - 2[0 \times (2 \; ft + 4 \; ft)]$

$= 20 \; ft$

The BLOS score for an LOS C falls between '2.5' and '3.5' per Exhibit 15-4. Based on this information, the effective speed factor S_t, and the posted speed limit S_p, shall be calculated using this range as follows:

Calculate speed for BLOS = 2.5:

$BLOS = 0.507 \ln(v_{OL})$
$\quad +0.1999 S_t (1 + 10.38 HV)^2$
$\quad +7.066(1/P)^2 - 0.005(W_e)^2 + 0.76$

$2.5 = 0.507 \ln(278)$
$\quad +0.1999 S_t (1 + 10.38 \times 0.03)^2$
$\quad +7.066(1/4)^2 - 0.005(20)^2 + 0.76$

$\to S_t = 1.35$

$S_t = 1.1199 \ln(S_p - 20) + 0.8103$

$\to S_p = 21.6 \, mph$

Calculate speed for BLOS = 3.5:

$BLOS = 0.507 \ln(v_{OL})$
$\quad +0.1999 S_t (1 + 10.38 HV)^2$
$\quad +7.066(1/P)^2 - 0.005(W_e)^2 + 0.76$

$3.5 = 0.507 \ln(278)$
$\quad +0.1999 S_t (1 + 10.38 \times 0.03)^2$
$\quad +7.066(1/4)^2 - 0.005(20)^2 + 0.76$

$\to S_t = 4.29$

$S_t = 1.1199 \ln(S_p - 20) + 0.8103$

$\to S_p = 42.9 \, mph$

Based on the above, in order to maintain an LOS C, speed shall be set between $21.6 \, mph$ and $42.9 \, mph$.

Correct Answer is (A)

SOLUTION 2.8
Chapter 18 Urban Street Segments, Section 4 Pedestrian Methodology of the *HCM Manual 6th edition* is used to solve this question. Steps 1 and 2 of Exhibit 18-17 are the only steps needed for this solution.

Step 1: Determine Free-Flow Walking Speed

The average Free-Flow walking speed $S_{pf} = 4.4 \, ft/sec$ when 0% to 20% of pedestrians travelling the segment are elderly.

Step 2: Determine Average Pedestrian Space

A. Compute Effective Sidewalk Width $'W_E'$

$W_E = W_T - W_{O,i} - W_{O,o} - W_{s,i} - W_{s,o}$

W_T is the sidewalk total width = $12 \, ft$

$W_{O,i}$ is the effective width of objects inside the sidewalk = 0

$W_{O,o}$ is the effective width of objects outside the sidewalk = 0

$W_{s,i}$ is the shy distance from the curb side and is calculated as follows were no buffer W_{buf} was provided in the question:

$W_{s,i} = max(W_{buf}, 1.5) = 1.5 \, ft$

$W_{s,o}$ is the shy distance on the outside (the none-curb side) and requires the proportion of window display $P_{window} = 300/1{,}000 = 0.3$, and the facade proportion $P_{building} = 0.7$ as inputs:

$W_{s,o} = 3.0 \, P_{window} + 2.0 \, P_{building}$
$\quad\quad +1.5 \, P_{fence}$
$\quad = 3.0 \times 0.3 + 2.0 \times 0.7 + 0$
$\quad = 2.3 \, ft$

$W_E = 12 - 0 - 0 - 1.5 - 2.3 = 8.2 \, ft$

B. Compute Pedestrian Flow Rate per Unit Width $'v_p'$

$$v_p = \frac{v_{ped}}{60\, W_E}$$

$$= \frac{4{,}000}{60 \times 8.2}$$

$$= 8.13\ p/ft/min$$

C. Compute Average Walking Speed $'s_p'$

$$S_p = (1 - 0.00078 v_p^2) S_{pf} \geq 0.5 S_{pf}$$

$$= (1 - 0.00078 \times 8.13^2) \times 4.4$$

$$= 4.17\ ft/sec$$

D. Compute Pedestrian Space $'A_p'$

$$A_p = 60\, \frac{S_p}{v_p} = 60 \times \frac{4.17}{8.13} \cong 31\ ft^2/p$$

Correct Answer is (B)

SOLUTION 2.9

Although the question can be answered with the use of some logic, will put a structure to it with the aim of visiting some of the important HCM chapters.

Chapter 18 Urban Street Segments, Section 4 Pedestrian Methodology is used to solve this question. Steps 5, 6, 7 and 8 are referred to in this solution.

Equation 18-38 that measures the difficulty factor F_{cd} is used for this assessment:

$$F_{cd} = 1 + \frac{0.10 d_{px} - (0.318\, I_{p,link} + 0.22\, I_{p,int} + 1.606)}{7.5}$$

I. *Increase in the number of through lanes in the subject direction.*

Reference is made to the following equations from step 6 Equation 18-32:

$$I_{p,link} = 6.0468 + F_w + F_v + F_s$$

$$F_v = 0.0091 \frac{v_m}{4\, N_{th}}$$

$$F_s = 4 \left(\frac{S_R}{100}\right)^2$$

The number of through lanes N_{th} is inversely proportional to the motorized vehicle volume adjustment factor F_v which in turn makes it inversely proportional to the pedestrian LOS score for link $I_{p,link}$. Overall, this makes N_{th} positively proportional to the difficulty factor F_{cd} due to the negative sign that precedes it in Equation 18-38.

This makes Statement I true.

II. *Increase in the distance from crossing point to the nearest signal-controlled crossing.*

d_{px} is the crossing delay in the difficulty factor equation and is directly proportional to d_{pd} the pedestrian diversion delay. This makes d_{pd} directly proportional to the difficulty factor.

This makes Statement II true.

III. *Increase in vehicles running speed.*

Motorized vehicle running speed S_R is directly proportional to the adjustment factor F_S – See Equation 18-35, also refer to the equations in statement I above. This makes it is directly proportional to the $I_{p,link}$.

Due to the negative sign that precedes the $I_{p,link}$ in Equation 18-38, this makes it inversely proportional to the difficulty factor.

This makes Statement III incorrect.

IV. *Increase in midsegment demand flow rate in the direction nearest to the sidewalk.*

Midsegment demand flow rate v_m is directly proportional to the motorized vehicle volume F_v which in turn makes it directly proportional to $I_{p,link}$, and in turn makes it inversely proportional F_{cd} as discussed previously.

This makes Statement IV incorrect.

Correct Answer is (A)

SOLUTION 2.10
Chapter 24 Off-Street Pedestrian and Bicycle Facilities, Step 2 of Exhibit 24-10 of the *HCM Manual 6th edition* is used to solve this question.

F_m is the number of meeting events and is calculated using Equation 24-6 as follows:

$$F_m = \frac{Q_{ob}}{PHF}\left(1 + \frac{S_p}{S_b}\right)$$

Where Q_{ob} is the bicycle demand in the opposite direction, S_p and S_b are the pedestrian and bicycle mean speeds respectively.

$$F_m = \frac{75}{0.85}\left(1 + \frac{4.3}{16}\right) = 112 \text{ events per hr } (*)$$

Correct Answer is (A)

(*) Although not requested in this question, the information above is not sufficient to determine *LOS* in case requested. In order to determine *LOS*, number of passing events shall be calculated as follows:

$$F_p = \frac{Q_{sb}}{PHF}\left(1 - \frac{S_p}{S_b}\right) = \frac{75}{0.85}\left(1 - \frac{4.3}{16}\right) = 64.5$$

The total number of events F which can be used to determine *LOS* is calculated as follows:

$$F = (F_p + 0.5 F_m)$$
$$= 64.5 + 0.5 \times 112$$
$$= 120.5 \text{ events per hr}$$

In reference to Exhibit 24-4, Level of Service in case requested is *LOS D* denoting *frequent conflicts with cyclists*.

SOLUTION 2.11
Chapter 24 Off-Street Pedestrian and Bicycle Facilities, Steps 1 to 5 of Exhibit 24-7 of the *HCM Manual 6th edition* are used to solve this question.

Step 1: Determine Effective Walkway Width

$$W_E = 5 \text{ ft}$$

Step 2: Calculate Pedestrian Flow Rate

$$v_{15} = \frac{v_h}{4 \times PHF} = \frac{1,200}{4 \times 0.83} \cong 361 \text{ p/h}$$

$$v_p = \frac{v_{15}}{15 \times W_E} = \frac{361}{15 \times 5} = 4.8 \text{ p/ft/min}$$

Step 3: Calculate Average Pedestrian Space

$$A_p = \frac{S_p}{v_p} = \frac{4 \text{ ft/s} \times 60 \text{ s/min}}{4.8 \text{ p/ft/min}} = 50 \text{ ft}^2/p$$

Step 4: Determine LOS

With the use of Exhibit 24-2, Level of Service with platoon-adjusted criteria for an $A_p = 50 ft^2/p$ is *LOS C*.

Step 5: Calculate Volume-to-Capacity Ratio

Capacity for walkways with platoon flow (average over 5 *min*) is $c = 18 \text{ p/min/ft}$

$$\frac{v}{c} = \frac{4.8 \text{ p/ft/min}}{18 \text{ p/min/ft}} = 0.3$$

Correct Answer is (D)

SOLUTION 2.12

Chapter 18 Urban Street Segments, Section 6 Transit Methodology of the *HCM Manual 6th edition* is used to solve this question. Steps 1, 2 and 3 of Exhibit 18-26 will be referred to in this solution.

Step 1: Determine Transit Vehicle Running Time

A. Compute Segment Running Speed $'S_{Rt}'$

$$S_{Rt} = \min\left(S_R, \frac{61}{1 + e^{-1.00 + (1,185 N_{ts}/L)}}\right)$$

Where S_R is the motorized vehicle running speed and is calculated as follows:

$$S_R = \frac{3,600\, L}{5,280\, T_R} = \frac{3,600 \times 2,000}{5,280 \times 40} = 34\ mph$$

Given that N_{ts} is the number of transit-stops on the segment which is one:

$$S_{Rt} = \min\left(34, \frac{61}{1 + e^{-1.00 + (1,185 \times 1/2,000)}}\right)$$

$$= \min(34, 36.6)$$

$$= 34\ mph$$

B. Compute Delay due to a Stop

1. Acceleration-deceleration delay $'d_{ad}'$

$$d_{ad} = \frac{5,280}{3,600}\left(\frac{S_{Rt}}{2}\right)\left(\frac{1}{r_{at}} + \frac{1}{r_{dt}}\right) f_{ad}$$

Where r_{at} and r_{dt} are acceleration and deceleration of transit vehicles taken as $3.3\ ft/s^2$ and $4.0\ ft/s^2$ respectively.

f_{ad} can be taken as '1.0' for stops not on the near side of boundary intersection.

$$= \frac{5,280}{3,600}\left(\frac{34}{2}\right)\left(\frac{1}{3.3} + \frac{1}{4.0}\right) \times 1.0$$

$$= 13.8\ s$$

2. Delay due to Serving Passengers $'d_{ps}'$

$$d_{ps} = t_d\, f_{dt}$$

Where t_d is the average dwell time given in the question as $25\ seconds$.

f_{dt} is the proportion of dwell time occurring during green. Given no signalized intersection identified $\rightarrow f_{dt} = 1.0$.

$$d_{ps} = 25 \times 1 = 25\ s$$

3. Reentry Delay $'d_{re}'$

Reentry delay is the time a transit vehicle waits for a gap to reenter a traffic stream. Given this transit passes through on-line stops, reentry is not applicable in this case $\rightarrow d_{re} = 0$.

4. Delay due to Stops $'d_{ts}'$

$$d_{ts} = d_{ad} + d_{ps} + d_{re}$$

$$= 13.8 + 25 + 0$$

$$= 38.8\ s$$

C. Compute Segment Running Time $'d_{Rt}'$

$$t_{Rt} = \frac{3,600\, L}{5,280\, S_{Rt}} + \sum_{i=1}^{N_{ts}} d_{ts,i}$$

$$= \frac{3,600 \times 2,000}{5,280 \times 34} + 38.8$$

$$= 78.9\ s$$

Step 2: Determine Delay at Intersection $'d_t'$

$$d_t = t_l\, 60 \left(\frac{L}{5{,}280}\right)$$

Given the description of the segment in the question → $d_t = 0$.

Step 3: Determine Travel Speed $'S_{Tl,seg}'$

$$S_{Tl,seg} = \frac{3{,}600\, L}{5{,}280(t_{Rt}+d_t)}$$

$$= \frac{3{,}600 \times 2{,}000}{5{,}280\,(78.9 + 0)}$$

$$= 17.3\ mph$$

Correct Answer is (C)

SOLUTION 2.13

The *NCEES Handbook*, Section 5.1.4.1 Acceleration of the Transportation chapter can be referred to for the solution of this question.

The given distance of 1.5 *mile* can be divided into three distances as follows:

1. *Acceleration distance* $'x_a'$

   $$x_a = \tfrac{1}{2} a t_a^2 + S_o t_a + x_o$$

   Where x_a refers to the distance being sought, t_a is the time needed to accelerate to speed S, and S_o is initial speed which is zero in this case.

   $$S = a\, t_a + S_o$$
   $$t_a = S/a = S/5 = 0.2\, S$$
   $$\to x_a = \tfrac{1}{2} a(S/a)^2 = 0.1\, S^2$$

2. *Deceleration distance* $'x_d'$

   $$x_d = \tfrac{1}{2} a t_d^2 + S_o t_o + x_o$$

   Assume that datum starts at the point of deceleration, and hence S_o and x_o will be taken as *zeros* and those shall be brought up in the final equation.

   $$S_{final} = a\, t_d + S = 0$$
   $$t_d = S/a = S/4.5 = 0.22\, S$$
   $$\to x_d = \tfrac{1}{2} a(S/a)^2 = 0.11\, S^2$$

3. *In-between (Middle) Distance*

   $$x_{mid} = S\, t_{mid}$$

Total distance $x = 1.5\ mile\ (= 7{,}920 ft)$:

$$x = x_a + x_d + x_{mid}$$
$$= 0.1 S^2 + 0.11 S^2 + S\, t_{mid}$$
$$= 0.21\, S^2 + S\, t_{mid}$$

Total time is 1.5 *minutes* (= 90*sec*):

$$90 = t_a + t_d + t_{mid}$$
$$= 0.42\, S + t_{mid}$$
$$\to t_{mid} = 90 - 0.42\, S$$

Substituting t_{mid} in the equation for $'x'$ above generates the following:

$$7{,}920 = 0.21 S^2 + S(90 - 0.42 S)$$
$$S^2 - 428.6\, S + 37{,}714.3 = 0$$

The above is a quadratic equation with $a = 1$, $b = -428.6$ and $c = 37{,}714.3$ and can be solved as follows:

$$root = \frac{-b \mp \sqrt{b^2 - 4ac}}{2a}$$

$$S = \frac{+428.6 \mp \sqrt{428.6^2 - 4 \times 1 \times 37{,}714.3}}{2 \times 1}$$

$$= (123.7, 304.9)$$

The more logical value for S is $123.7 ft/sec\ (= 84.3 mph)$

Correct Answer is (B)

SOLUTION 2.14

Chapter 19 Signalized Intersections, Exhibit 19-18, *Step 4 Determine Adjusted Saturation Flow Rate* of the *HCM Manual 6th edition* is used to solve this question.

Given this lane is an exclusive lane and information on pedestrians or bicycles were not provided, the following equation can be used:

$$s = s_o f_w f_{HVg} f_p f_{bb} f_a f_{LU} f_{LT} f_{RT} f_{Lpb} f_{Rpb} f_{wz} f_{ms} f_{sp}$$

s_o is the base saturation rate and is taken from Exhibit 19-11 for metropolitan population with $pop. > 250,000$ as $s_o = 1,900 \ pc/h/ln$

f_w is the adjustment factor for lane width and is taken as '1.0' for lane width $= 10 \ ft$

f_{HVg} is the adjustment factor for Heavy Vehicles and Grade and is calculated as follows:

$$f_{HVg} = \frac{100 - 0.78 \, p_{HV} - 0.31 \, P_g^2}{100}$$

$$= \frac{100 - 0.78 \times 7 - 0}{100}$$

$$= 0.95$$

f_p is the adjustment factor for parking when vehicles are permitted to park adjacent to the lane being analyzed. This factor is calculated as follows:

$$f_p = \frac{N - 0.1 - \frac{18 \, N_m}{3,600}}{N} \geq 0.050$$

$$= \frac{1 - 0.1 - \frac{18 \times 10}{3,600}}{1}$$

$$= 0.85$$

f_a is the adjustment factor for the area type and is taken as '0.9' for CBD areas.

f_{LU} is the adjustment factor for Lane Utilization $= 1.0$.

Given the above adjustment factors, the adjusted saturation rate is calculated as follows:

$$s = s_o f_w f_{HVg} f_p f_a f_{LU}$$

$$= 1,900 \times 1.0 \times 0.95 \times 0.85 \times 0.9 \times 1.0$$

$$= 1,381 \ veh/h$$

Capacity for a one traffic movement with no permitted left of right lanes is calculated using Equation 19-16:

$$c = N \, s \, \frac{g}{C}$$

$$= 1.0 \times 1,381 \times \frac{20}{100}$$

$$= 276 \ veh/h \quad (*)$$

Correct Answer is (A)

(*) Although not requested in this question, it is to be noted that every lane group's capacity is calculated differently.

Relevant equations and capacity equations for left-turn and right-turn lane groups, exclusive or shared, permitted, protected, or protected-permitted, and others, can be obtained from the *HCM Manual* Chapter 31 Equations 31-117 to 31-129 of the 7th edition – available online.

SOLUTION 2.15

The following chapters from the HCM manual are used to solve this question:

- Chapter 19 Signalized Intersections, Exhibit 19-8, *Step 4 Determine Adjusted Saturation Flow Rate.*
- Chapter 31 Signalized Intersections: Supplemental, Section 3, *Step 6 Determine Lane Group Saturation Flow Rate (page 31-51).*

Saturation flow rate for a permitted right-turn operation in a shared lane is calculated using Equation 31-105:

$$S = \frac{S_{th}}{1 + P_R \left(\frac{E_R}{f_{Rpb}} - 1\right)}$$

S_{th} is the saturation flow rate of the exclusive through lane which is calculated as follows:

$$s_{th} = s_o f_w f_{HVg} f_p f_{bb} f_a f_{wz} f_{ms} f_{sp}$$

s_o is the base saturation rate and is taken from Exhibit 19-11 for metropolitan population with $pop. > 250,000$ as $s_o = 1,900 \ pc/h/ln$

f_w is the adjustment factor for lane width and is '0.96' for lane width $< 10 \ ft$

f_{HVg} is the adjustment factor for Heavy Vehicles and Grade and is calculated as follows for negative grades:

$$f_{HVg} = \frac{100 - 0.79 \, p_{HV} - 2.07 \, P_g}{100}$$

$$= \frac{100 - 0.79 \times 10 - 2.07(-2)}{100}$$

$$= 0.96$$

f_a is the adjustment factor for the area type and is taken as '0.9' for CBD areas.

The adjusted saturation rate for the exclusive lane s_{th} is defined in one of Equation 31-105 variables page 31-52:

$$s_{th} = s_o f_w f_{HVg} f_a$$

$$= 1,900 \times 0.96 \times 0.96 \times 0.9$$

$$= 1,576 \ veh/h/ln$$

P_R is the proportion of right-turning vehicles in the shared lane and is calculated as follows:

$$P_R = \frac{v_{Turning \ right}}{v_{Total}} = \frac{100}{400} = 0.25$$

E_R is the equivalent number of through cars for a protected right-turning vehicle and can be taken as '1.18'.

f_{Rpb} is the pedestrian-bicycle adjustment factor for right-run groups and is calculated using Chapter 31 of the *HCM Manual 7th edition* Equations 31-74 to 31-83 as follows:

a. Determine Pedestrian Flow Rate During Service

$$v_{pedg} = v_{ped} \frac{C}{g_{ped}} \leq 5,000$$

$$= 75 \times \frac{120}{10}$$

$$= 900 \ p/h$$

b. Determine Average Pedestrian Occupancy

$$OCC_{pedg} = \frac{v_{pedg}}{2,000}$$

$$= \frac{900}{2,000}$$

$$= 0.45$$

c. Determine Bicycle Flow Rate During Green – Not used.

d. Determine Average Bicycle Occupancy – Not used.

e. Determine Relevant Conflict Zone Occupancy

$$OCC_r = \frac{g_{ped}}{g} OCC_{pedg}$$

$$= \frac{10}{20} \times 0.45$$

$$= 0.225$$

f. Determine Unoccupied Time

Given there is only one cross-street receiving lane, the following more conservative equation is used:

$$A_{pbT} = 1 - OCC_r$$

$$= 1 - 0.225$$
$$= 0.775$$

g. Determine Saturation Flow Rate Adjustment Factor

For permitted right-turn operation in an exclusive lane:

$$f_{Rpb} = A_{pbT} = 0.775$$

Provided the above calculated adjustment factors, saturation rate can now be determined as follows:

$$S = \frac{S_{th}}{1+P_R\left(\frac{E_R}{f_{Rpb}}-1\right)}$$

$$= \frac{1,576}{1+0.25\left(\frac{1.18}{0.775}-1\right)}$$

$$= 1,394 \; veh/h/ln$$

Correct Answer is (B)

SOLUTION 2.16

Chapter 31 Signalized Intersections: Supplemental, of the *HCM Manual 6th edition* is used to solve this question.

Saturation flow rate for a permitted left-turn operation lane is calculated using Equation 31-100 (*):

$$S_p = \frac{v_o e^{-v_o t_{cg}/3,600}}{1 - e^{-v_o t_{fh}/3,600}}$$

v_o is the opposing through demand flow rate which equals to 550 veh/h given that the opposing right-turn vehicles have their own lane hence would not interfere with the "left-turners" of the lane being analyzed.

t_{cg} is the critical headway, and t_{fh} is the follow-up headway, and both can be taken as 4.5 *Seconds* and 2.5 *Seconds* respectively.

$$S_p = \frac{v_o e^{-v_o t_{cg}/3,600}}{1 - e^{-v_o t_{fh}/3,600}}$$

$$= \frac{550 \times e^{-550 \times 4.5/3,600}}{1 - e^{-550 \times 2.5/3,600}}$$

$$= 871 \; veh/h/ln$$

Correct Answer is (A)

(*) Refer to Equations 31-110 & 111 for an exclusive left-turn in case requested in the exam.

SOLUTION 2.17

Establishing a queuing diagram is the best way to solve this question.

A queuing diagram has the number of vehicles represented by its y-axis and time represented by its x-axis copied here for ease of reference:

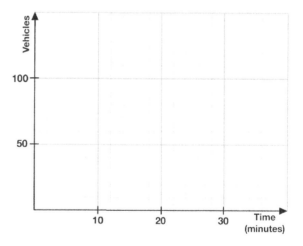

There are three events which have occurred and that can be plotted on this diagram using the following equations:

Original rate of arrival 5 veh/min:

$$V_{arrival} = 5t \quad \text{referred to as } \lambda \text{ in literature}$$

Delayed rate of departure 2 veh/min:

$$V_{delayed} = 2t \quad (t \leq 20min)$$
$$\text{referred to as } u \text{ in literature}$$

Facilitated rate of departure 11 veh/min:

$V_{facilitated} = 11t \quad (t > 20 min)$

The time the maximum queue dissipates takes place when the rate of arrival equals to the rate of facilitated departure considering the duration it took to clear out the accident:

$V_{arrival}|_{t=0}^{t=t} = V_{delayed}|_{t=0}^{t=20} + V_{facilitated}|_{t=20}^{t=t}$

$[5(t) - 5(0)] = [2(20) - 2(0)] +$
$\qquad\qquad\qquad\qquad [11(t) - 11(20)]$

$5t = 2(20) + 11t - 220$

$\rightarrow t = 30 \; minutes$

The time the maximum queue (Q_{max}) occurs is at $t = 20$ as shown in the following diagrams. It therefore takes Q_{max} to dissipate:

$30 - 20 = 10 \; minutes$

The above three equations can be plotted using the following diagram:

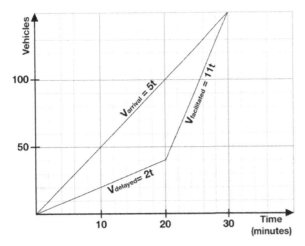

The maximum queue which would occur is represented by the maximum vertical distance shown on the diagram below (Q_{max}) which takes place at time $t = 20 \; min$.

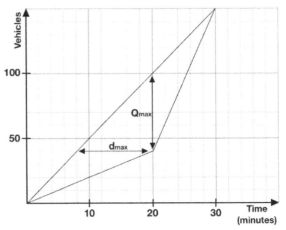

Q_{max} can either be measured – provided a to-scale diagram, or calculated as follows:

$Q_{max} = V_{arrival(@\,t=20)} - V_{delayed(@\,t=20)}$

$\qquad = 5(20) - 2(20)$

$\qquad = 60 \; vehicles$

Correct Answer is (C)

Question Extras:

Calculating Maximum Delay
d_{max} on the second diagram represents the maximum delay experienced by drivers which equals to $12 \; minutes$ in this question as measured from the diagram.

Calculating Total Delay
Total delay experienced by all vehicles is represented by the area inside the queuing diagram.

This area can be measured using a to-scale diagram, trigonometry, or can be calculated using integration as follows:

$= \int_0^{30} 5t \cdot dt - \int_0^{20} 2t \cdot dt - \int_0^{10}(40 + 11t) \cdot dt$

$= \frac{5}{2}t^2 \Big|_{t=0}^{t=30} - \frac{2}{2}t^2 \Big|_{t=0}^{t=20} - (40t + \frac{11}{2}t^2)\Big|_{t=0}^{t=10}$

$= \frac{5}{2}(30)^2 - (20)^2 - (40(10) + \frac{11}{2}(10)^2)$

$= 900 \; veh.min$

SOLUTION 2.18

Chapter 13 Freeway Weaving Segments of the *HCM Manual 6th edition* is used to solve this question.

The following steps shall be performed to obtain all required information needed to substitute in the *capacity* equation:

Determine the number of lanes from which weaving maneuvers may be completed with either one or no lane changes N_{WL}:

This information can be determined by observing the provided weaving segment diagram given in the question.

Vehicles moving on the freeway must (1) change one lane to join the weaving segment, and (2) vice versa, which makes $N_{WL} = 2$

Determine the Volume Ratio VR, which is the flow of weaving vehicles to the total flow:

$$VR = \frac{v_w}{v_{total}} = \frac{150+850}{700+150+850+50} = 0.57$$

Determine the maximum weaving length L_{max} to confirm that this is a weave segment not a merge/diverge segment using Equation 13-4:

$$L_{max} = [5{,}728(1 + VR)^{1.6}] - (1{,}566\, N_{WL})$$
$$= [5{,}728(1 + 0.57)^{1.6}] - (1{,}566 \times 2)$$
$$= 8{,}656\, ft > 2{,}000\, ft \rightarrow ok$$

This information could be obtained from Exhibit 13-11 as well. This result confirms that the solution can proceed with using equations provided in Chapter 13.

Determine the capacity of the freeway using Exhibit 12-6 as follows:

$$c_{IFL} = 2{,}200 + 10 \times (FFS - 50)$$
$$= 2{,}200 + 10 \times (60 - 50)$$
$$= 2{,}300\, pcphpl$$

Determine capacity controlled by Density C_{IWL} using Equation 13-5:

$$c_{IWL} = c_{IFL} - [438.2(1 + VR)^{1.6}] + (0.0765\, L_S) + (119.8 N_{WL})$$
$$= 2{,}300 - [438.2(1 + 0.57)^{1.6}] + (0.0765 \times 2{,}000) + (119.8 \times 2)$$
$$= 1{,}791\, pcphpl$$

Determine capacity controlled by weaving demand flows c_{IW} for N_{WL} using Equation 13-7 as follows:

$$C_{IW} = \frac{2{,}400}{0.57} = 4{,}215\, veh/h$$

Divide by # of lanes = $1{,}405\, pcphpl$

The lowest of those two capacities control, in which case $c = 1{,}405\, pcphpl$

Correct Answer is (A)

SOLUTION 2.19

Chapter 13 Freeway Weaving Segments of the *HCM Manual 6th edition* is used to solve this question.

The following steps shall be performed to obtain all required information needed to substitute in the *weaving intensity factor* Equation 13-20.

Determine the number of lanes from which weaving maneuvers may be completed with either one or no lane changes N_{WL}:

This information can be determined by observing the provided weaving segment with the Managed Laned diagram provided in the question.

Vehicles moving on the freeway must (1) change one lane to join the Managed Lane, and (2) vice versa, which makes $N_{WL} = 2$

Determine the Volume Ratio VR, which is the flow of weaving vehicles to the total flow:

$$VR = \frac{v_W}{v_{total}} = \frac{750+500}{1,700+750+500+200} = 0.40$$

Determine the maximum weaving length L_{max} to confirm that this is a weave segment not a merge/diverge segment using Exhibit 13-11:

Using Exhibit 13-11 with $N_{WL} = 2$ and $VR = 0.4 \rightarrow L_{max} = 6,681 \, ft$. This result confirms that the solution can proceed using equations provided in Chapter 13.

L_{max} can be obtained by using Equation 13-4 as well as demonstrated in Solution 2.18.

Determine lane changing rates using Equations 13-11 to 13-17 as follows:

Weaving vehicles change rate:

$$LC_W = LC_{MIN} + 0.39[(L_S - 300)^{0.5} N^2 (1 + ID)^{0.8}]$$

$$LC_{MIN} = (LC_{RF} \times v_{RF}) + (LC_{FR} \times v_{FR})$$

LC_{RF} & LC_{FR} are the minimum number of lane changes from the *Ramp* (i.e., the Managed Lane) to the *Freeway* and vice versa, in which case, and by observing the weaving diagram, both values equal to '1.0'. v_{RF} & v_{FR} are the volume from the *Managed Lane* to the *Freeway*, and vice versa. N is the number of lanes in which case this equals to '3' and ID is the interchange density within 3 miles upstream and 3 miles downstream of the center of the weaving segment divided by six – in which case $ID = 0$.

$$LC_{MIN} = (1 \times 500) + (1 \times 750)$$
$$= 1,250 \, lcph$$

$$LC_W = 1,250 + 0.39[(1,000 - 300)^{0.5} 3^2 (1 + 0)^{0.8}]$$
$$= 1,343 \, lcph$$

Non-weaving vehicles change rate Eq. 13-12:

$$I_{NW} = \frac{L_S \times ID \times v_{NW}}{10,000} = \frac{1,000 \times 0 \times (1,700+200)}{10,000} = 0$$

Provided that $I_{NW} < 1,300$, Equation 13-13 is used in this case:

LC_{NW}
$$= (0.206 v_{NW}) + (0.542 L_S) - (192.6 N)$$
$$= (0.206 \times 1,900) + (0.542 \times 1,000) - (192.6 \times 3)$$
$$= 356 \, lcph$$

The above result is true as long as Equation 13-16 is satisfied, and the above value is $< 2,135 + 0.223(v_{NW} - 2000)$
$< 2,135 + 0.223(1,900 - 2000)$
$< 2,113 \rightarrow OK$

Total lane-changing rate Eq. 13-17:

$$LC_{ALL} = LC_W + LC_{NW}$$
$$= 1,343 + 356$$
$$= 1,699 \, lcph$$

Weaving intensity factor Eq. 13-20:

$$W = 0.226 \left(\frac{LC_{ALL}}{L_S}\right)^{0.789}$$
$$= 0.226 \left(\frac{1,699}{1,000}\right)^{0.789}$$
$$= 0.34$$

Correct Answer is (B)

Question Extras:

Average speed of weaving vehicles Eq. 13-19:

$$S_W = 15 + \left(\frac{FFS \times SAF - 15}{1+W}\right)$$

$$= 15 + \left(\frac{65 \times 1 - 15}{1 + 0.34}\right)$$

$$= 52.3 \, mph$$

No adverse factors given in the question that could affect Speed → $SAF = 1.0$

Average speed of non-weaving vehicles Eq. 13-21:

$$S_{NW} = FFS \times SAF - (0.0072LC_{MIN}) - \left(0.0048\frac{v}{N}\right)$$

$$= 65 \times 1 - (0.0072 \times 1{,}250) - \left(0.0048 \times \frac{1{,}700 + 750 + 500 + 200}{3}\right)$$

$$= 51 \, mph$$

Average speed of all Vehicles Eq. 13-22:

$$S = \frac{v_W + v_{NW}}{\left(\frac{v_W}{S_W}\right) + \left(\frac{v_{NW}}{S_{NW}}\right)}$$

$$= \frac{1{,}250 + 1{,}900}{\left(\frac{1{,}250}{52.3}\right) + \left(\frac{1{,}900}{51}\right)}$$

$$= 51.5 \, mph$$

Density D and LOS Eq. 13-23:

$$D = \frac{(v/N)}{S}$$

$$= \frac{(3{,}150/3)}{51.5}$$

$$= 20.4 \, pcpmpl$$

Using Exhibit 13-6, a density of $20.4 \, pcpmpl$ in a freeway weaving segment is assigned a Level of Service $LOS \, C$.

SOLUTION 2.20

Chapter 19 Signalized Intersections of the *HCM Manual 6th edition* is used to solve this question.

It is important to understand some terminologies, abbreviations and references which are explained in Exhibit 19-29 for minor crossings and Exhibit 19-30 for major crossings.

Given the crossing in this question is for a major street, Exhibit 19-30 is used along with all its identifiers which are as follow:

d references a major crossing when the major crossing is the subject of analysis.

c references a minor crossing adjacent to the major crossing d.

o references outbound pedestrians, which refers to pedestrians crossing from the side we choose to analyze.

i references inbound pedestrians, which refers to pedestrians crossing from the other side.

v_{do} refers to the outbound flow of pedestrians and (d) in this case references a major crossing as indicated above.

v_{di} refers to the inbound flow of pedestrians on the major crossing.

v_{co} references pedestrians flow joining from crossing the minor street next to the major crossing. This flow is not used in this question since it is not required in any of the equations.

With the above in mind, the following steps shall be performed to obtain the required information needed to substitute in the *crosswalk occupancy time* Equation 19-67:

OUTBOUD Ped. (assume left to right)

Determine the number of pedestrians arriving during each cycle to cross the major street using Equation 19-53:

$$N_{do} = \frac{v_{do}}{3{,}600} C$$

$$= \frac{700}{3{,}600} \times 110$$

$$= 21.4\ p$$

Determine the effective walk time for the minor street $g_{walk,mi}$ which equals to the walk setting phase $Walk_{mi}$ time plus 4.0 sec per Equation 19-54 for pretimed signals:

$$\rightarrow g_{walk,mi} = 14\ sec$$

Calculate pedestrian service time arriving at the corner to cross the major street for crosswalks wider than 10 ft using Equation 19-64:

$$T_{ps,do} = 3.2 + \frac{L_d}{S_p} + 2.7 \frac{N_{ped,do}}{W_d}$$

Where $N_{ped,do}$ is the number of pedestrians who will cross when the sign indicates so and is calculated per Equation 19-66 as follows:

$$N_{ped,do} = N_{do} \frac{C - g_{walk,mi}}{C}$$

$$= 21.4 \times \frac{110 - 14}{110}$$

$$= 18.7\ p$$

$$T_{ps,do} = 3.2 + \frac{L_d}{S_p} + 2.7 \frac{N_{ped,do}}{W_d}$$

$$= 3.2 + \frac{48}{4} + 2.7 \frac{18.7}{14}$$

$$= 18.8\ s$$

INBOUD Ped. (assume right to left)

$$N_{di} = \frac{v_{di}}{3{,}600} C$$

$$= \frac{650}{3{,}600} \times 110$$

$$= 19.9\ p$$

$$N_{ped,di} = N_{di} \frac{C - g_{walk,mi}}{C}$$

$$= 19.9 \times \frac{110 - 14}{110}$$

$$= 17.4\ p$$

$$T_{ps,di} = 3.2 + \frac{L_d}{S_p} + 2.7 \frac{N_{ped,di}}{W_d}$$

$$= 3.2 + \frac{48}{4} + 2.7 \frac{17.4}{14}$$

$$= 18.6\ s$$

The *crosswalk occupancy time* is therefore calculated using Equation 19-67 as follows:

$$T_{occ} = T_{ps,do} N_{do} + T_{ps,di} N_{di}$$

$$= 18.8 \times 21.4 + 18.6 \times 19.9$$

$$= 772.5\ p.s$$

Correct Answer is (A)

SOLUTION 2.21
Chapter 14 Freeway Merge and Diverge Segments of the *HCM Manual 6th edition* is used to solve this question.

Exhibit 14-6 is referred to in this chapter, few steps can be omitted though provided some of the information are given in the questions. Step 1 can be omitted in this case.

Flow rates are to be calculated for lanes 1 and 2 'v_{12}' per Step 2 for an on-ramp merge segment as follows:

$$v_{12} = v_f \times P_{FM}$$

v_f is the freeway flow rate upstream of the on-ramp – add to it flow from *ramp 1*.

P_{FM} is the proportion of freeway vehicles remaining in lanes 1 and 2 immediately upstream of the on-ramp influence area. Adjacent on-ramps have no statistical significance on the operation and hence can be ignored, and for that, Equation 14-3 can be used in this case:

$$P_{FM} = 0.5775 + 0.000028\, L_A$$

L_A is the length of the acceleration lane and is calculated using Exhibit 14-5 ($= 890\, ft$).

$P_{FM} = 0.5775 + 0.000028 \times 890 = 0.6$

$v_{12} = v_f \times P_{FM}$
$= (4{,}000 + 350) \times 0.6$
$= 2{,}610\, pcph$

Check if lane 3 has a flow rate $> 2{,}700\, pcph$ or $> 1.5 \times \left(\frac{v_{12}}{2}\right)$ – if so, v_{12} needs to be adjusted per Equations 14-15 or 16:

$v_3 = v_f - v_{12}$
$= (4{,}000 + 350) - 2{,}610$
$= 1{,}740\, pcph < 2{,}700\, ok$

$\qquad < 1.5 \times \left(\frac{v_{12}}{2}\right) = 1{,}957\, ok$

$\to v_{12}$ need not to be adjusted $= 2{,}610\, pcph$

Density in on-ramp (merge) influence area is calculated using Equation 14-22 as follows:

$D_R = 5.5475 + 0.00734 v_R$
$\qquad\qquad + 0.0078 v_{12} - 0.00627 L_A$

$= 5.5475 + 0.00734 \times 350$
$\qquad + 0.0078 \times 2{,}610 - 0.00627 \times 890$

$= 23\, pcpmpl$

Correct Answer is (B)

SOLUTION 2.22

Chapter 22 Roundabouts of the *HCM Manual 6th edition* is used to solve this question.

Exhibit 22-10 steps are followed to solve this question. Steps 1 and 2 can be ignored given that $PHF = 1.0$ and $f_{HV} = 1.0$, hence demand volume (V) in veh/h is equivalent to flow rates (v) in $pcph$.

$$v_{i,pce} = \frac{V}{PHF \times f_{HV}}$$

Step 3: Determine circulating and exiting flow rates for the bypass lane

In order to determine the capacity of the bypass, the conflicting traffic to the bypass is taken into account, in which case, this is the exit just next to it (Eastbound direction) as shown in the below figure:

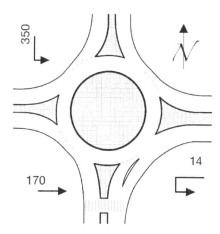

In which case, the exiting flow rate through the Eastbound direction is:

$v_{ex,EB,pce} = 350 + 170 + 14$
$\qquad\quad\; = 534\, pcph$

Step 4: Determine entry flow rates by lane – for the bypass lane

This is given in the question as:

$v_{bypass,EB} = 400\, pcph$

Step 5: Determine capacity of the bypass lane

Use Equation 22-6 to calculate the capacity for a bypass opposed by one exiting lane as follows:

$$c_{bypass,EB,pce} = 1{,}380 e^{(-1.02 \times 10^{-3}) v_{ex,EB,pce}}$$
$$= 1{,}380 e^{(-1.02 \times 10^{-3}) \times 534}$$
$$= 800 \ pcph$$

Step 6: Determine pedestrian impedance to vehicles

Use Exhibit 22-18 to calculate pedestrians' adjustment factor f_{ped} for $n_{ped} \leq 101$ as follows:

$$f_{ped} = 1 - 0.000137 n_{ped}$$
$$= 1 - 0.000137 \times 75$$
$$= 0.99$$

Step 7: Convert bypass lane flow rate and capacity into veh/h

Given that $PHF = 1.0$ and $f_{HV} = 1.0$:

$$V_{bypass,EB} = 400 \ veh/h$$

Step 8: Compute the volume-to-capacity ratio for the bypass lane

Using Equations 22-16, volume to capacity is calculated as follows with due respect to correcting capacity to account for impeding pedestrians:

$$x_{bypass,EB} = \frac{v_{bypass,EB}}{c_{bypass,EB,pce} \times f_{ped}}$$
$$= \frac{400}{800 \times 0.99}$$
$$= 0.51$$

Correct Answer is (A)

SOLUTION 2.23

Chapter 20 Two-Way Stop-Controlled Intersections of the *HCM Manual 6th edition* is used to solve this question.

Exhibit 20-6 is referred to and in specific steps 1 to 5a for non-signalized intersections.

Step 1: Determine and label movement priorities

The below diagram labels movements copied from Exhibit 20-1 for ease of reference:

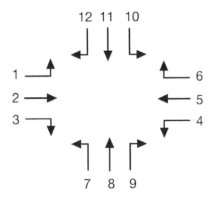

Four-Leg Intersection

Movement 1 is what this question seeks and is identified as rank 2 major street left turn. This movement is conflicted by movements 5 and 6 – check Exhibit 20-7 and Equation 20-2 for more details. Given that pedestrians' flow is negligible, movement 16 (identified in the said Exhibit) is set to zero.

Step 2: Convert movement demands into flow rates

$$v_i = \frac{V_i}{PHF}$$

$v_1 = \frac{70}{0.9} = 78\ veh/h$

$v_5 = \frac{170}{0.9} = 189\ veh/h$

$v_6 = \frac{45}{0.9} = 50\ veh/h$

Step 3: Determine conflicting flow rates

Equation 20-2 is used in this case:

$v_{c,1} = v_5 + v_6 + v_{16}$

$= 189 + 50 + 0$

$= 239\ veh/h$

Step 4: Determine critical headways and follow-up headway

Critical headway for the movement, which represents an acceptable gap, is calculated using Equation 20-30 and Exhibit 20-12 as follows:

$t_{c,1} = t_{c,base} + t_{c,HV}\ P_{HV} + t_{c,G}G - t_{3,LT}$

$= 4.1 + 0 + 0 - 0$

$= 4.1\ s$

Follow-up headway for the movement is calculated using Equation 20-31 and Exhibit 20-13 as follows:

$t_{f,1} = t_{f,base} + t_{f,HV}\ P_{HV}$

$= 2.2 + 0$

$= 2.2\ s$

Step 5a: Compute potential capacities

$c_{p,1} = v_{c,1} \frac{e^{-v_{c,1}t_{c,1}/3{,}600}}{1-e^{-v_{c,1}t_{f,1}/3{,}600}}$

$= 239 \frac{e^{-239 \times 4.1/3{,}600}}{1-e^{-239 \times 2.2/3{,}600}}$

$= 1{,}400\ veh/h$

Correct Answer is (B)

SOLUTION 2.24

Chapter 5 Transportation of the *NCEES Handbook*, Section 5.4.1 Dilemma Zones is used to solve this question.

Dilemma zone X_{dz} is calculated as follows:

$X_{dz} = X_c - X_o$

In order to determine the required speed to avoid this zone from forming, dilemma zone is set to equal zero. The resultant is a quadratic equation as follows:

$X_c = X_o$

$1.47V(t_{stop}) + 1.075 \frac{V^2}{a_2}$

$= 1.47VY - W + \frac{1}{2}\ a_1(Y - t_{passing})^2$

$1.47 \times 1.5\ V + 1.075 \frac{V^2}{10}$

$= 1.47 \times 4V - (40 + 14) + \frac{1}{2} \times 10(4 - 1.5)^2$

$V^2 - 34.2\ V\ + 211.6 = 0$

The above is a quadratic equation with $a = 1$, $b = -34.2$ and $c = 211.6$ and can be solved as follows:

$root = \frac{-b \mp \sqrt{b^2 - 4ac}}{2a}$

$V = \frac{+34.2 \mp \sqrt{34.2^2 - 4 \times 1 \times 211.6}}{2 \times 1}$

$= (8\ mph, 26\ mph)$

A few trials can be conducted in order to check whether the dilemma zone is truly eliminated between those two velocities.

Correct Answer is (A)

SOLUTION 2.25

Chapter 23 Ramp Terminals and Alternative Intersections of the *HCM Manual 6th edition* is used to solve this question.

I. DDIs have freeway entry and exit ramps separated at the street level creating four intersections.

Section 2, Concepts, of the same chapter is referred to.

Dimond Interchanges sub-Section defines Split Diamond Interchanges as those ones which have freeway entry and exit ramps separated at street level. A sketch is also provided in Exhibit 23-15. This however does not apply to DDIs. See Exhibit 23-16.

For the removal of doubt, one can search up internet animation videos for each interchange movements for better understanding.

This marks Statement I as incorrect.

II. Large efficiency gains for interchanges with heavy left-turn demand.

Section 2, Concepts, of the same chapter is referred to.

Diverging Dimond Interchanges DDI sub-section states that DDIs allow free flowing left-turns onto the freeway.

This makes Statement II true.

III. Some unique considerations need to take place for lane utilization and saturation flow rate adjustments for crossovers and U-turn movements.

The first part of the statement "...*unique considerations need to take place for lane utilization and saturation flow rate adjustments ...*" can be vetted against the same sub-section Diverging Dimond Interchanges DDI and is true.

The second portion of the statement can be checked against Exhibit 23-4 of the same chapter where it mentions that DDIs **DO NOT** require adjustments for saturation flow rate for U-turns.

This marks Statement III as incorrect.

IV. DDIs allow free flowing left turn movements onto the crossing freeway.

Section 2, Concepts, of the same chapter is referred to.

Diverging Dimond Interchanges DDI sub-section claims that DDIs result in large efficiency gains for interchanges with heavy left-turn demands.

This makes Statement IV true.

Correct Answer is (D)

SOLUTION 2.26

Chapter 4C Traffic Control Signal Needs Studies and Chapter 8C Flashing-Light Signals, Gates, and Traffic Control Signals of the *MUTCD Manual* are used to solve this question.

Traffic signal warrants 1, 2 and 9 are checked against traffic to understand if a signal is warranted, after which, Chapter 8C is consulted for the rail automatic gate.

The two-direction volume for the major approach are added and used along with the corresponding largest one-direction traffic volume of the minor approach, based upon which the traffic data is reorganized as follows:

Facility	Major Rd Traffic	Minor Rd. Max traffic	Rail
8AM – 9AM	374	98	2
9AM – 10AM	297	102	
10AM – 11AM	260	82	
11AM – 12AM	203	56	
12AM – 1PM	340	102	
1PM – 2PM	273	36	2
2PM – 3PM	186	56	
3PM – 4PM	180	97	
4PM – 5PM	316	102	2

Warrant 1: Eight-Hour Vehicular Volume

Checking with the maximum traffic volume in the reorganized table, 374 veh/h, against Table 4C-1 Condition A of the manual under column 70%, given that the posted speed is > 40 mph, is lesser than the minimum required of 420 veh/h. Same applies to the minor street. Based upon which this warrant condition is **NOT satisfied**.

The same process can be applied using Table 4C-1 Condition B under 70%, and same conclusion applies.

Warrant 2: Four-Hour Vehicular Volume

Using Figure 4C-2, a plot of the maximum traffic volume for the major road along with the corresponding minor road traffic fall below the curve. Based on this, this warrant is **NOT satisfied** as well.

Warrant 9: Intersection Near a Crossing Grade

In this warrant, the need for a signal is justified if the nearest grade center line is within 140 ft of the street stop line. Figure 4C-9 is then used along with the modification factors for the minor traffic volume per Tables 4C-2, 3 and 4.

Trigonometry is applied to understand if the nearest track centerline is within 140 ft of the minor street stop line as follows:

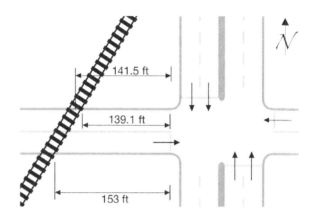

As observed from this figure, the track center line falls outside the prescribed 140 ft. Based upon which a traffic control signal is **NOT warranted**.

Finally, and in reference to Chapter 8C of the manual, page 773, Section 8C.05 Clauses 01 and 02 state that an automatic gate is required, even if speed is 25 mph, as long as the grade is at an intersection.

As a conclusion, traffic signal is NOT warranted for this intersection, however, an automatic gate for the rail is required.

The above is a high-level conclusion and an engineering study of traffic conditions will be required to confirm it.

Correct Answer is (C)

SOLUTION 2.27

Chapter 6H Typical Applications of the *MUTCD Manual* is used to solve this question.

Typical Application TA 30 page 692 & 693 of the manual deals with temporary traffic control measures for low-speed low volume urban streets with one interior lane closed.

In reference to the figure shown in TA 30 and to Table 6H-3 of the manual:

$$d3 = A = 100 \, ft$$

$$d4 = B = 100 \, ft$$

Taper length $'L'$ is calculated using Table 6H-4 of the manual as follows:

$$d2 = L = \frac{WS^2}{60} = \frac{12 \times 40^2}{60} = 320 \, ft$$

The buffer space $d1$ is optional, it can be a zero or can be determined using Table 6C-2 of the manual as follows:

$$d1 = Buffer\ space = 305 \, ft$$

Correct Answer is (A)

SOLUTION 2.28

Chapter 2C Warning Signs and Object Markers of the *MUTCD Manual* is used to solve this question.

Table 2C-5 Horizontal Alignment Sign Selection specifies that chevron signs (W1-8) are required at speeds of 15 *mph* which rules out option (A).

Table 2C-6 of the same chapter specifies that sign spacing is 40 *ft* for the given radius and speed.

Correct Answer is (B)

SOLUTION 2.29

Chapter 5 Transportation of the *NCEES Handbook,* Section 5.4.2 Offsets, can be referred to for the solution of this question.

In order to determine the required cycle length C_{prog}, travel speed V is required, which can be determined using the given offset as follows:

$$offset = \frac{d_o}{V}$$

$$C_{prog} = \frac{d_o}{V} \times 2$$

$$V = \frac{d_o}{offset}$$

$$= \frac{800 \, ft}{10 \, sec}$$

$$= 80 \, ft/sec \, (= 54 \, mph)$$

In order to have a good progression in both directions, the cycle length needs to be twice the travel time from intersection No. 1 and intersection No. 2, and hence, intersection No. 3 is not taken into account in this equation. See below:

$$C_{prog} = \frac{d_o}{V} \times 2$$

$$= \frac{800 \, ft}{80 \, ft/sec} \times 2$$

$$= 20 \, sec$$

Correct Answer is (D)

SOLUTION 2.30

Chapter 2H General Information Signs, and Chapter 4C Traffic Control Signal Needs Studies are used to solve this question.

I. For traffic control signals when coordinated, a sign should be used to identify the coordinated speed.

Section 4C.07 Warrant 6, Coordinated Signal System, says that progressive movement sometimes necessitates the installation of such a sign, and this makes the use of this sign not mandatory. Section 2H.03 Clause 01 confirms this as well.

This marks Statement I as incorrect.

II. Signs used to identify a coordinated signals system can have a changeable message for speed and time if those were not consistent during the day/week.

Section 2H.03 Traffic Signal Speed Sign (I1-1), clause 02 states that a changeable message can be used if progression speeds are set for different times of the day.

This makes Statement II true.

III. They are installed predominantly for one way direction streets to provide the necessary degree of vehicular platooning.

Section 4C.07 Warrant 6, Coordinated Signal System, requires a study to confirm the need for such a system, the study is normally triggered by the need mentioned in this statement not vice versa.

This marks Statement III as incorrect.

IV. They are best applied when the resultant spacing between signals is less than 1,000 ft

The Section 4C.07 Warrant 6, states that the system should NOT be applied if the resultant spacing is less than 1,000 ft.

This marks Statement III as incorrect.

Correct Answer is (C)

SOLUTION 2.31

Chapter 19 Signalized Intersections and Chapter 31 Signalized Intersections of the *HCM Manual 6th edition* are used to solve this question.

Uniform delay is calculated using Equation 19-19 as follows:

$$d_1 = PF \frac{0.5\, C(1 - g/C)^2}{1 - [min(1, X)\, g/C]}$$

Where C is cycle length, g is the effective green. PF is the progression adjustment factor, this can either be computed using Equation 19-21 or collected from Chapter 31 Exhibit 31-38 using the information given in this question as '0.7'.

X is the volume-to-capacity ratio and is calculated using both Equations 19-16 and 19-17 as follows:

$$X = \frac{v}{c} \quad \& \quad c = N\, s\, \frac{g}{C}$$

$$X = \frac{v}{N\, s\, \frac{g}{C}}$$

$$= \frac{300}{1 \times 1,750 \times \frac{20}{100}}$$

$$= 0.86$$

$$d_1 = PF \frac{0.5\, C(1-g/C)^2}{1-[min(1,X)\, g/C]}$$

$$= 0.7 \times \frac{0.5 \times 100\, (1-20/100)^2}{1-[min(1, 0.86) \times 20/100]}$$

$$= 27\ sec$$

Correct Answer is (B)

SOLUTION 2.32

Chapter 2C Warning Signs and Object Markers, and Chapter 3B Pavement and Curb Marking of the *MUTCD Manual* is used to solve this question.

Distance $d1$ is taken from Figure 3B-14 represented by L as follows:

$$d1 = L = WS = 12 \times 60 = 720 \, ft$$

Distances $d2$ and $d3$ are represented by d of Figure 3B-14 and they are taken from Table 2C-4 for condition A "heavy traffic" and a speed of 60 *mph* as follows:

$$d = d2 + d3 = 1{,}100 \, ft$$

$d3$ represents the end of road marking indicating a lane reduction ahead equals to quarter of d as follows:

$$d3 = 1{,}100/4 = 275 \, ft$$

$$\rightarrow d2 = 1{,}100 - 275 = 825 \, ft$$

Correct Answer is (A)

SOLUTION 2.33

Chapter 4 Network Screening of the *HSM Manual 1st edition*, Section 4.4.2.2 Crash Rate, is used to solve this question.

Crash rates are calculated and ranked in the respective table as follows – crash rate calculation is shown below for intersection No. 1 only:

$$Crash \, Rate \, (R_i) = \frac{N_{Observed,i}}{MEV_i}$$

Where $N_{Observed,i}$ is the total observed crashes at intersection 'i' and MEV_i is the million entering vehicles at intersection 'i'.

$$MEV_1 = \frac{17{,}800 + 12{,}100}{1{,}000{,}000} \times 5 \times 365 = 54.6$$

$$R_1 = \frac{97}{54.6} = 1.8$$

Intersection	MEV_i	Crashes	R_i
1	54.6	97	1.8
2	44.9	46	1.02
3	51.6	30	0.58
4	28.3	21	0.74

Based on the above, the ranking from the highest rate to the lowest rate would be as follows: 1, 2, 4, 3

Correct Answer is (A)

SOLUTION 2.34

Chapter 4 Network Screening of the *HSM Manual 1st edition*, Section 4.4.2.3 Equivalent Property Damage Only (EPDO) Average Crash Frequency, is used to solve this question.

The EPDO method assigns weighing factors using the societal costs in reference to the PDO cost and multiply them with the number of crashes to determine the PDO score.

Weighing factors are presented in the following table with the first one solved for demonstration as shown below:

$$f_{i \, (weight)} = \frac{CC_i}{CC_{PDO}}$$

Crash severity	Comprehensive Crash Unit Cost	$f_{i \, (weight)}$
Fatal (K)	$5,740,100	568
Disability (A)	$304,400	30
Evident (B)	$111,200	11
Possible (C)	$62,700	6
PDO (O)	$10,100	1

$$f_{K\,(weight)} = \frac{CC_k}{CC_{PDO}}$$

$$= \frac{\$5{,}740{,}100}{\$10{,}100}$$

$$= 568$$

$$PDO\ Score = \sum f_i \times N_{observed}$$

$$= 568 \times 3 + 30 \times 4 + 11 \times 13$$

$$+ 6 \times 33 + 1 \times 47$$

$$= 2{,}212$$

Correct Answer is (D)

SOLUTION 2.35
Chapter 13 Roadway Segments of the *HSM Manual 1st edition*, Section 13.5.2 Roadside Element Treatments with CMFs, is used to solve this question.

Table 13-18 of the above section is used which assigns CMF_{Total} for this improvement a value of '0.85'.

$$Reduction = 55 - 55 \times 0.85$$

$$= 8.25\ crash/yr$$

Correct Answer is (A)

SOLUTION 2.36
Chapter 6 Select Countermeasures of the *HSM Manual 1st edition* can be referred to provide a solution for this question.

The above chapter offers numerous measures and contributing factors for consideration to reduce accidents.

Section 6.2.2 of the same chapter specifies that vehicles' rollover is mainly caused by inadequate shoulder width. **Which renders statement I as correct.**

Other measures, or contributing factors, could help in reducing rollovers, but those have not been identified as the main cause for rollovers.

Correct Answer is (B)

SOLUTION 2.37
The solution is achieved by multiplying and then adding the two tables together making sure that zero household owning vehicles are contributing to the bottom line as well as those trips weren't only specified for vehicle owning household. See below:

Overall trips generated	Vehicles ownership		
Household size	0	1	2 +
1	300	600	1,200
2	675	900	2,000
3	1,400	2,125	2,700
4	2,000	2,730	3,850
5 +	3,120	4,550	5,800
Sub Total	**7,495**	**10,905**	**15,550**

Adding all trips together generates the following:

$$= 7{,}495 + 10{,}905 + 15{,}550$$

$$= 33{,}950\ trips\ per\ day$$

Correct Answer is (A)

SOLUTION 2.38
A quick method can be applied by taking advantage of the units provided as follows:

$$Density(veh/mile) = \frac{flow\ rate\ (veh/hr)}{av.travel speed\ (mph)}$$

$$= \frac{1{,}500\ veh/hr}{65\ mile/hr}$$

$$= 23\ veh/mile$$

Another method of solving this problem can be found in Chapter 4 of the *HCM Manual* – this is presented here so you can benefit from the method of deriving other values:

$$flow\ rate\ (veh/hr) = \frac{3{,}600\ sec/hr}{av.\ headway\ (sec/veh)}$$

$$\rightarrow Av.\ headway\ (sec/veh) = \frac{3{,}600\ sec/hr}{Flow\ rate\ (veh/hr)}$$

$$= \frac{3{,}600\ sec/hr}{1{,}500\ veh/hr}$$

$$= 2.4\ sec/veh$$

$$Av.\ headway\ (sec/veh) = \frac{av.\ spacing\ (ft/veh)}{av.\ travel\ speed\ (ft/sec)}$$

$$2.4\ s/veh = \frac{av.\ spacing\ (ft/veh)}{65\ mph \times \frac{5{,}280\ ft/mile}{3{,}600\ sec/hr}}$$

$$\rightarrow Av.\ spacing\ (ft/veh)$$

$$= 2.4\ sec/veh \times 65\ mph \times \frac{5{,}280\ ft/mile}{3{,}600\ sec/hr}$$

$$= 228.8\ ft/veh$$

$$Density\ (veh/mile) = \frac{5{,}280\ ft/mile}{av.\ spacing\ (ft/veh)}$$

$$= \frac{5{,}280\ ft/mile}{228.8\ ft/veh}$$

$$= 23\ veh/mile$$

Correct Answer is (D)

SOLUTION 2.39

In the *NCEES Handbook version 2.0* is used to solve this problem, Section 5.1.3.3 Peak-Hour Factor as follows:

$$PHF_{SW-NE} = \frac{V}{V_{15} \times 4}$$

Where V represents the four consecutive 15 minutes period that give us the highest value – see below table. V_{15} represents the highest flow within that hour.

In the following table, X represents V_{15} at 4:45 PM :

Time	SW-NE direction Veh/hr	NE-SW direction Veh/hr
4:00 PM – 4:45 PM	761	540 + X
4:15 PM – 5:00 PM	763	**542 + X**
4:30 PM – 5:15 PM	**778**	496 + X
4:45 PM – 5:30 PM	777	456 + X

$$PHF_{SW-NE} = \frac{778}{210 \times 4} = 0.92$$

$$PHF_{NE-SW} = \frac{0.92}{1.05} = 0.88$$

At this stage, and in order to calculate PHF, $V_{15,max}$ could either be 197 veh or $'X'$ (only if $X > 197$). Both values will be used to determine $'X'$ as follows:

Taking $V_{15} = 197$:
$$PHF_{NE-SW} = \frac{542 + X}{197 \times 4} = 0.88$$

$$\rightarrow X = 151$$

Taking $V_{15} = X$:
$$PHF_{NE-SW} = \frac{542 + X}{X \times 4} = 0.88$$

$$\rightarrow X = 215$$

Although mathematically the two $'X'$ values can work, applying logic and observing the data around the given hour, a value of **215 *veh*** within that hour seems more logical compared to a value of **151 *veh***.

Correct Answer is (B)

SOLUTION 2.40

The *NCEES Handbook version 2.0*, Section 5.1.4.1 Acceleration, of the Transportation chapter can be referred to solve this question.

In this question we shall determine the total distance traveled x_{total} and divide it by the total time it took the transit to travel this distance t_{total}, as follows:

$$x_{total} = x_a + x_c + x_d$$

$$t_{total} = t_a + t_c + t_d$$

$$S_{average} = \frac{x_{total}}{t_{total}}$$

1. Acceleration distance 'x_a'

$$x_a = \tfrac{1}{2} a t_a^2 + S_o t_a + x_o$$

Where x_a is the distance being sought, t_a is the time needed to accelerate to speed $S = 90 \, mph$ calculated as follows where S_o is initial speed taken is zero in this case:

$$S = a\, t_a + S_o$$

$$90 \, mph \times \frac{5{,}280 \, ft/mile}{3{,}600 \, sec/hr} = 5.0 \, \tfrac{ft}{sec^2} \times t_a + 0$$

$$t_a = 26.4 \, sec$$

$$x_a = \tfrac{1}{2} a t_a^2 + 0 + 0$$

$$= \tfrac{1}{2} (5.0 \, \tfrac{ft}{sec^2})(26.4 \, sec)^2$$

$$= 1{,}742.4 \, ft$$

2. Deceleration distance 'x_d'

$$x_d = \tfrac{1}{2} a t_d^2 + S_o t_o + x_o$$

Will assume that datum starts at the point of deceleration, and hence S_o and x_o will be taken as zeros and those shall be brought up in the final equation.

$$S_{final} = a\, t_d + S = 0$$

$$(-4.5 \, \tfrac{ft}{sec^2}) t_d + 90 \, mph \times \frac{5{,}280 \, ft/mile}{3{,}600 \, sec/hr} = 0$$

$$t_d = 29.33 \, sec$$

$$x_d = \tfrac{1}{2} a t_d^2 + 0 + 0$$

$$= \tfrac{1}{2} (4.5 \, \tfrac{ft}{sec^2})(29.33 \, sec)^2$$

$$= 1{,}936 \, ft$$

3. In-between (constant) distance 'x_c'

$$x_c = S\, t_c$$

$$= \left(90 \, mph \times \frac{5{,}280 \, ft/mile}{3{,}600 \, sec/hr}\right)\left(1.5 \, min \times 60 \, \tfrac{sec}{min}\right)$$

$$= 11{,}880 \, ft$$

$$x_{total} = x_a + x_c + x_d$$

$$= 1{,}742.4 + 11{,}880 + 1{,}936$$

$$= 15{,}558.4 \, ft \, (2.95 \, mile)$$

$$t_{total} = t_a + t_c + t_d$$

$$= 26.4 + 1.5 \times 60 + 29.33$$

$$= 145.73 \, sec \, (0.0405 \, hr)$$

$$S_{average} = \frac{2.95 \, mile}{0.0405 \, hr} = 72.8 \, mph$$

Correct Answer is (C)

SOLUTION 2.41

The *NCEES Handbook*, Chapter 5 Transportation, can be referred to for the skid marks equation stopping distance in Section 5.1.4.3 as follows:

$$d_b = \frac{v_1^2 - v_2^2}{30(f \mp G)}$$

For the first car travelling at a speed of 90 mph, the stopping distance is calculated as follows:

$$d_{b,1} = \frac{v_1^2 - 0}{30(f \mp 0)} = \frac{270}{f}$$

Distance traveled by the first car during the perception/reaction time of the driver:

$d_{p,1} = v \times t$

$= 90 \text{ mph} \times \frac{5{,}280 \text{ ft/mile}}{3{,}600 \text{ sec/hr}} \times 2 \text{ sec}$

$= 264 \text{ ft}$

For the second car travelling at a speed of 60 mph, the stopping distance is calculated as follows:

$$d_{b,2} = \frac{v_1^2 - 0}{30(f \mp 0)} = \frac{120}{f}$$

Distance traveled by the second car during the perception/reaction time of the driver:

$d_{p,1} = v \times t$

$= 60 \text{ mph} \times \frac{5{,}280 \text{ ft/mile}}{3{,}600 \text{ sec/hr}} \times 2 \text{ sec}$

$= 176 \text{ ft}$

For those two cars not to hit each other, and keep a distance of 35 ft between them after stopping, the following equation applies:

$d_{b,1} + d_{b,2} + d_{p,1} + d_{p,2} + 35 \text{ ft} = 1{,}340 \text{ ft}$

$\frac{270}{f} + \frac{120}{f} + 264 + 176 + 35 = 1{,}340$

$\rightarrow f = 0.451$

Reapply the friction factor to the initial equations to determine both $d_{b,1}$ and $d_{b,2}$:

$$d_{b,1} = \frac{270}{f} \approx 599 \text{ ft}$$

$$d_{b,2} = \frac{120}{f} \approx 266 \text{ ft}$$

Correct Answer is (B)

SOLUTION 2.42

Start by rearranging all speeds from least to greatest as follows:

SN	Speed mph
1	12
2	21
3	25
4	32
5	45
6	45
7	45
8	46
9	55
10	56
11	59
12	59
13	69
14	72
15	73

The 85th percentile speed is the speed at or below which 85% of all vehicles are observed travelling past a monitoring point.

The position where 85% of the dataset fall under equals to $0.85 \times 15 = 12.75$. Since this is not a whole number, the whole number right after this position will be used, in which case '13'. In other means, 85% of the speeds observed are below the speed of position 13 which is 69 mph - this speed represents the 85th percentile of this data set.

Correct Answer is (A)

SOLUTION 2.43

The *NCEES Handbook version 2.0*, Chapter 5 Transportation, Section 5.4.3.1 Yellow Change Interval is used in this solution.

$$Y = t + \frac{v}{2a + 2Gg}$$

With no input given for time or acceleration, $'t'$ can be taken as 1 second and $'a'$ is typically 10 ft/sec^2 as stated in the *NCEES Handbook*.

Also, note that speed in this formula is in ft/sec.

$$Y = 1 + \frac{1.47 \frac{ft/sec}{mph} \times 45\ mph}{2 \times 10\ ft/sec^2 + 2 \times 0 \times 32.2\ ft/sec^2}$$

$$= 4.3\ seconds$$

Correct Answer is (B)

SOLUTION 2.44
Chapter 6F Toll Road Signs of the *MUTCD Manual* is referred to in this question.

Table 2F-1 defines toll road sign and plaque sizes. In case of an auxiliary sign, identified as M4-15 in this table, which is a sign placed on top of route numbers (see Section 2F.11 page 242 for more details), its minimum size when used in conventional roads should be 24 $in \times$ 12 in.

Correct Answer is (A)

SOLUTION 2.45
Chapter 4C, Section 4C.06 Warrant 5, School Crossing, of the *MUTCD Manual* is referred to in the solution.

Warrant 5 states that a traffic engineering study is required to assess the situation further, moreover Section 4C.01 states that "the satisfaction of a traffic warrant or warrants shall not in itself require the installation of a traffic control signal". Which indicates that a traffic engineering study is required in all cases.

This makes option (A) valid.

The school crossing warrant represented by warrant 5 defines schoolchildren as those children in elementary and high school (almost everyone) and that there should be a minimum number of 20 schoolchildren present or crossing during peak hour.

This makes option (B) valid.

The warrant requires the frequency of gaps to be studied in combination with the period when schoolchildren are using the crossing.

The provided statement in the question is incomplete in this sense and hence does not qualify as a prerequisite or condition for this warrant.

This makes option (C) invalid.

The location of a traffic signal and its proximity to the school crossing can factor into the decision of installing a signal for the crossing. For instance, if the distance between the school crossing and an intersection with a traffic signal is less than 300 ft, this eliminates the requirement behind this warrant and the existing traffic control system can be adjusted and made coordinated instead to account for schoolchildren crossing it.

This makes option (D) valid.

Correct Answer is (A + B + D)

PART II
Traffic Engineering

DESIGN & GEOMETRY

Knowledge Areas Covered

SN	Knowledge Area
3	**Roadside and Cross-Section Design** A. Forgiving roadside concepts (e.g., clear zone, recoverable slopes, roadside obstacles) B. Barrier design (e.g., barrier types, end treatments, crash cushions) C. Cross-section elements (e.g., lane widths, shoulders, bike lane, sidewalks, retaining walls) D. Nonmotorized design considerations (e.g., shared-use paths, bicycle facilities, pedestrian facilities, ADA compliance, traffic-calming features)
4	**Horizontal Design** A. Basic circular curve elements (e.g., middle ordinate, length, chord definition, radius definition, centerline stationing) B. Sight distance considerations C. Superelevation (e.g., rate, transitions, method, components) D. Special horizontal curves (e.g., compound/reverse curves, curve widening, coordination with vertical geometry)
5	**Vertical Design** A. Vertical alignment (e.g., geometrics, vertical clearance) B. Stopping and passing sight distance (e.g., crest curve, sag curve)
6	**Intersection Geometry** A. Intersection sight distance B. Interchanges (e.g., freeway merge, entrance and exit design, horizontal design, vertical design) C. At-grade intersection layout, including roundabouts

PART III
Design & Geometry

PROBLEM 3.1 *Basic Horizontal Curve*

The below is a plan view for an 11 ft wide road with a horizontal curve passing through its centerline and an obstruction as shown. The setback from this obstruction to the edge of the road should not be less than 12 ft.

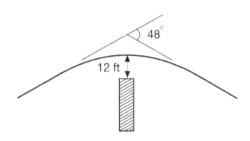

Based on this, the radius, and the length of the curve in ft should nearly be:

(A) 138 & 116

(B) 36 & 31

(C) 53 & 44

(D) 200 & 170

PROBLEM 3.2 *Basic Vertical Curve*

The below vertical curve has a 14 ft bridge passing over the curve's PVI as shown:

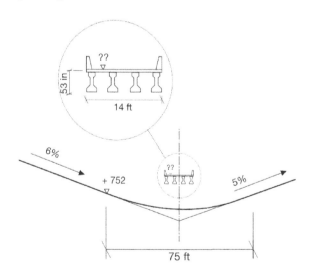

With a 16 ft clearance, and asphalt level at PVC of +752, bridge deck level should be designed at:

(A) + 750.9

(B) + 771.3

(C) + 771.0

(D) + 769.9

PROBLEM 3.3 *Points on Vertical Curve*

The following points fall on a vertical curve profile:

Point	Station	Elevation
PVC	0 + 025	72.5
PVI	0 + 275	65
A point on the curve	0 + 150	70

Using the above data, the initial and final grade for this section are as follows:

(A) −0.03 , 0.05

(B) −0.03 , 0.06

(C) −0.06 , 0.10

(D) −0.06 , 0.08

PROBLEM 3.4 *Double Horizontal Curve*

The below figure depicts two identical horizontal curves connecting two parallel roads as shown with station $St.1 = 1 + 00$

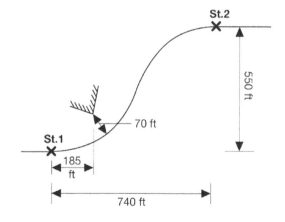

An obstruction is located at the middle ordinate of the bottom curve with a distance of $M = 70\ ft$.

Provided the above, station $St.2$ most nearly sits at:

(A) $8 + 40$

(B) $10 + 38$

(C) $10 + 78$

(D) $11 + 18$

PROBLEM 3.5 *Compound Horizontal Curve*

Use the following bearings for a compound horizontal curve:

- Back tangent $N\ 45°\ 45'\ E$
- Common tangent $S\ 80°\ 30'\ E$
- Forward tangent $S\ 41°\ 45'\ E$

The above tangents are connected with a $1{,}200\ ft$ long compound curve.

Given that the two curves making the compound curve are equal, the ratio of the flatter curve's radius to the sharper curve's radius is most nearly:

(A) $2 : 1$

(B) $1.1 : 1$

(C) $1 : 1$

(D) $1.4 : 1$

PROBLEM 3.6 *Addition of a Compound Horizontal Curve*

In the following diagram, the solid curve represents an existing road, while the dotted curve represents an additional new road:

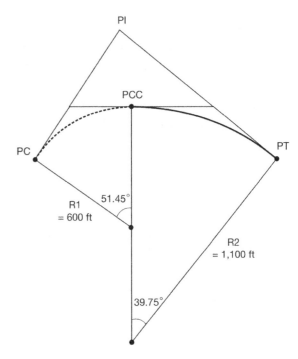

Given that the station at PI is $3 + 00$, the station at point PCC would most nearly be:

(A) $15 + 67.3$

(B) $4 + 00$

(C) $0 + 12.5$

(D) $1 + 10.4$

PROBLEM 3.7 *Superelevation and Road Cross Section – Spiral Curve*

A two-lane highway at a horizontal alignment, with each lane $12\ ft$ wide, transitions from a normal crowned cross-section of 2% to a fully superelevated cross section with $e = 8\%$ using a $250\ ft$ long spiral curve (*).

Provided that the superelevation is attained by revolving the road cross section about its centerline, the following cross section best represents that of the entire road midway through the spiral curve:

(A) Cross Section A

(B) Cross Section B

(C) Cross Section C

(D) Cross Section D

(*) Assume that the superelevation runoff is accomplished over the length of the spiral curve as recommended by the Green Book Section 3.3.8.4.6.

PROBLEM 3.8 *Basic Superelevation*
The superelevation of a two-lane highway horizontal curve that has a radius of 900 ft, and each lane is 12 ft wide, is being designed.

Considering the design speed for the curve is 50 mph, the outer edge of the pavement with respect to the inner edge is most nearly raised by:

(A) 13 in

(B) 6.5 in

(C) 22 in

(D) 26 in

PROBLEM 3.9 *Superelevation Transition*
With a maximum superelevation of 8% that should be attained at the Point of Curve (PC) by revolving the pavement cross-section around the centerline, along with a design speed of 50 mph and a radius for the horizontal curve of 2,500 ft, the designer decided to allocate portion of the superelevation runoff to the straight tangent in advance of the PC for safety reasons.

The longitudinal gradient for the 12 ft wide one lane cross-section's edge of pavement should not exceed 0.5% (or a slope of 1: 200) to maintain the best driving experience.

Given the above information, the maximum length of superelevation runoff that can be assigned to the straight tangent is most nearly:

(A) 192 ft

(B) 103 ft

(C) 307 ft

(D) 153 ft

PROBLEM 3.10 *Superelevation and Road Cross Section – Straight Tangent*

A two-lane highway at a horizontal alignment, with each lane 12 ft wide, transitions from a normal crowned cross-section of 2% to a fully superelevated cross-section with $e = 6\%$ at the Point of Curve (PC) using a straight tangent for an approach.

Given that the superelevation is attained by revolving the road cross-section about its inner edge, along with a radius of horizontal curve of 700 ft and a design speed of 50 mph, the following cross section best represents that of the road 50 ft ahead of the PC:

(A) Cross Section A

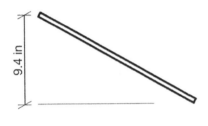

(B) Cross Section B

(C) Cross Section C

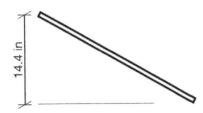

(D) Cross Section D

PROBLEM 3.11 *Reverse Horizontal Curve*

The perpendicular distance between two tangents is 55 ft. Those tangents are connected using a revere horizontal curve with a central or deflection angle (for both) of 30° degrees.

If the radius for one of the curves is made equal to 75 ft, the total arc length for the two connected curves is most nearly:

(A) 83 ft

(B) 215 ft

(C) 210 ft

(D) 80 ft

PROBLEM 3.12 *Curve Widening*

A four-lane highway is designed to accommodate a WB-92D (Rocky Mountain Double-Trailer Combination) truck with a design speed of 45 mph, and a 12 ft wide lane.

As the highway encounters a horizontal curve with a radius of 1,000 ft, the traveled way widening that should take place at this curve is most nearly:

(A) 2.6 ft

(B) 3.0 ft

(C) 6.0 ft

(D) 5.2 ft

PROBLEM 3.13 *Horizontal Sight Distance*

A one-lane ramp with a lane width of 12 ft, a 4 ft shoulder and another 4 ft wide ditch, is on a horizontal curve with a radius of 900 ft and a posted speed limit of 35 mph.

The horizontal curve has an object placed 5 ft away from the outer edge of its ditch which is one of the reasons why the speed is limited as such. The ramp is also located on a downgrade of 6%.

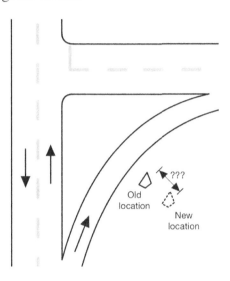

The authorities are considering moving this object away from the ramp in a bid to increase its speed limit by another 15 mph.

The minimum distance this object should be moved away from its original location as shown is most nearly:

(A) 31 ft

(B) 10 ft

(C) 12 ft

(D) 17 ft

(✽) PROBLEM 3.14 *Horizontal Curve Coordinate System (1)*

The below figure represents a roadway that has a horizontal curve starting at 2 + 35 at its Point of Curve PC along with a deflection angle of 37.5° as shown.

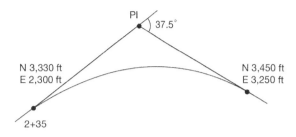

Given the above information, the Northing and Easting coordinates for station 9 + 35 are most nearly:

(A) N 3,480 ft, E 2,977 ft

(B) N 4,022 ft, E 2,977 ft

(C) N 3,537 ft, E 3,235 ft

(D) N 3,480 ft, E 3,135 ft

(✽) PROBLEM 3.15 *Horizontal Curve Coordinate System (2)*

The below figure represents a roadway that has a horizontal curve starting at 1 + 00 at its Point of Curve PC along with a deflection angle of 35° as shown.

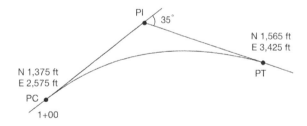

Given the above information, the Northing and Easting coordinates for station 11 + 50 are most nearly:

(A) N 1,551 ft, E 3,590 ft

(B) N 1,589 ft, E 3,590 ft

(C) N 1,400 ft, E 3,439 ft

(D) N 1,361 ft, E 2,590 ft

PROBLEM 3.16 *Spiral Curve Length*
The appropriate and most desirable length for a spiral curve that enhances drivers' experiences ahead of a horizontal circular curve that has a radius of 900 ft and a design speed of 65 mph is most nearly:

(A) 240 ft

(B) 120 ft

(C) 267 ft

(D) 191 ft

PROBLEM 3.17 *Vertical Curve Length*
The minimum length for a vertical curve that satisfies SSD and that has an approach grade of −7% and a departure grade of 4% along with a design speed of 65 mph is most nearly:

(A) 1,727 ft

(B) 471 ft

(C) 1,099 ft

(D) 4,752 ft

PROBLEM 3.18 *Escape Ramp*
The below WB-92D truck is running out of control at a speed of 100 mph headed into an escape ramp made of Pea Gravel. The truck managed to stop at distance L.

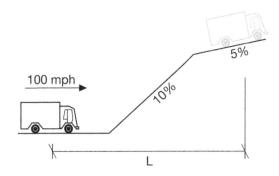

Given that the first slope is 500 ft long with an upgrade of 10%, and the second slope has an upgrade of 5%, distance L is most nearly:

(A) 1,030 ft

(B) 910 ft

(C) 1,530 ft

(D) 795 ft

PROBLEM 3.19 *Vertical Curve Design with a Bridge on Top (1)*
The below is a symmetrical parabolic vertical curve that shall maintain a 16 ft clearance to the bottom of the bridge as shown.

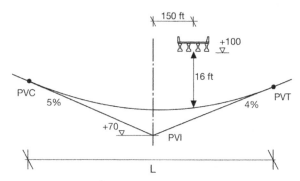

Given the above information, the length of the vertical curve L is most nearly:

(A) 793 ft

(B) 1,238 ft

(C) 2,451 ft

(D) 1,803 ft

PROBLEM 3.20 *Vertical Curve Design with a Bridge on Top (2)*
The below vertical curve is to be designed for trucks with a speed of 65 mph assuming the following:

- Deceleration rate of 11.2 ft/sec^2
- Eye height for truck drivers is 8 ft
- Object height that represents vehicles' taillights is 2 ft
- Use only SSD for sight distance if needed.

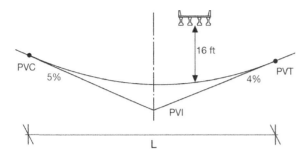

The recommended minimum length L that should satisfy the curve's sight distance, given that the bridge minimum clearance is 16 ft as shown above, is most nearly:

(A) 795 ft

(B) 425 ft

(C) 645 ft

(D) 311 ft

PROBLEM 3.21 *Crest Curve Slope*
The below crest curve starts at a PVC of 1 + 00 with an upward going slope of (+4%). The PVI is located at 5 + 00 with a level of +100 after which it heads into a downward slope of (−3%).

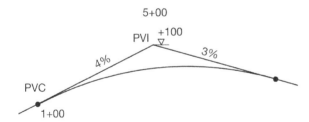

Given the above information, a tangent sloped with (−1.5%) gradient is located at the following station:

(A) 6 + 35

(B) 4 + 14

(C) 7 + 29

(D) 8 + 40

PROBLEM 3.22 *Crest Curve Design Speed*
A 1,000 ft long crest curve with a 3% upward tangent and a 3% downgrade tangent is the subject of study to determine its safest design speed.

Considering the height of a normal driver's eye is 3.5 ft and an object that represents the taillight of vehicles ahead is 2 ft, drivers' reaction time of 2.5 sec along with a deceleration rate of 11.2 ft/sec^2, the safest design speed for this curve should nearly be:

(A) 54 mph

(B) 62 mph

(C) 70 mph

(D) 76 mph

PROBLEM 3.23 *Crest Curve PSD*
The minimum length required for a crest curve with an upgoing grade of 3% and a down going grade of 3% that satisfies the Passing Sight Distance PSD for normal passengers' vehicles with a design speed of 60 mph is most nearly:

(A) 2,142 ft

(B) 1,000 ft

(C) 570 ft

(D) 816 ft

(⁂) PROBLEM 3.24 *Crest Vertical Curve Passing a Fixed Point*
The below vertical curve must achieve a level of +87.5 at station 2 + 73 to service the entrance of a facility that is located there.

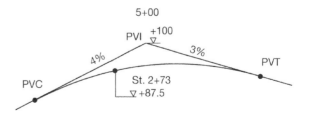

Given the station and the level at PVI are 5 + 00 and +100 respectively, the curve Length L should nearly be:

(A) 185 ft

(B) 1,241 ft

(C) 1,114 ft

(D) 1,253 ft

PROBLEM 3.25 *Lowest Point on Sag Curve*

The below sag curve's PVC starts at station 1 + 20 as shown below, elevation at PVI is +77 with a downgrade of 4% and an upgrade of 3%.

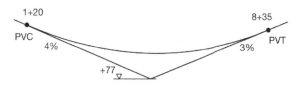

Given the above information, the elevation of the lowest point on this curve is most nearly:

(A) +83

(B) +82

(C) +80

(D) +85

PROBLEM 3.26 *Horizontal and Vertical Curve Coordination*

The following statements are true when it comes to coordinating vertical and horizontal curves during design:

I. Horizontal curves should not begin close to the top of a crest curve.
II. The vertical curve should lead the horizontal curve.
III. It is best to locate the flat portion of the vertical profile near the flat cross section of the superelevation transition.
IV. Sharp horizontal curves should not be introduced near the bottom of steep grades approaching a sag curve.

(A) II + III + IV

(B) II + III

(C) I + II + III

(D) I + IV

PROBLEM 3.27 *Parking Lane Considerations*

To prevent motorists in commercial areas from using the parking lane as part of a right-turn movement, and to eliminate inefficiencies of such an operation, or to prevent vehicles from mounting the curb and accidently hitting a pole when turning right, it is recommended to end the parking lane in advance of the intersection by at least:

(A) 10 ft

(B) 20 ft

(C) 30 ft

(D) 40 ft

PROBLEM 3.28 *Local Road Maximum Grade*

A local road is being designed to serve a rural mountainous area with a design speed of 25 mph.

The maximum grade this road can be designed for is:

(A) 5%

(B) 6%

(C) 10%

(D) 15%

PROBLEM 3.29 *Collector Road Minimum Roadway Width*

A collector road that serves a community that has many farms and agricultural land is being assessed for width. The volume of traffic expected is nearly 200 *veh/day* and design speed should not exceed 45 *mph*.

Given the above information, the minimum total roadway width (with shoulders) that should be considered for this collector is:

(A) 18 *ft*

(B) 20 *ft*

(C) 22 *ft*

(D) 24 *ft*

PROBLEM 3.30 *Superelevation Treatment in a Divided Arterial Road*

The best method for attaining superelevation for a 70 *ft* wide divided four-lane arterial road located in an urban area with two lanes in each direction, each 12 *ft* wide, with a shoulder width of 4 *ft*, and a speed limit of 60 *mph* is the following:

(A) The median between the two roads is held in a horizontal plane and the two roads are rotated separately around the median edges.

(B) Each of the two roads are treated on their own with no regard to the median.

(C) The entire roadway (i.e., the four lanes, shoulders and median) are treated as one plane and they are superelevated altogether.

(D) No superelevation is required for arterials located in urban areas.

PROBLEM 3.31 *Freeway Median Width*

The preferred minimum median width for a six-lane freeway – three lanes per direction – with a truck directional design hourly volume of 300 *veh/h* located in an urban area is:

(A) There are no freeways in urban areas.

(B) 22 *ft*

(C) 26 *ft*

(D) 32 *ft*

(⁂) PROBLEM 3.32 *Intersection Sight Distance Case B*

Passenger car A waits at a stop sign on a minor road with its rear wheels on a 5% upgrade. The major road has four 12 *ft* wide lanes and a 6 *ft* wide median. The Major road is on a 5% upgrade toward its north direction.

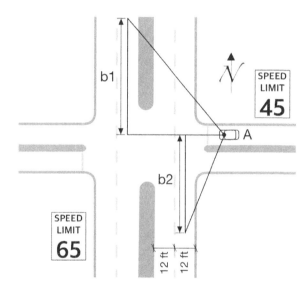

For a left turn, the Intersection Sight Distance ISD $b1$ should be _____ *ft*, and for a right turn, $b2$ should be _____ *ft*.

(⁂) Normally you are asked to provide one value.

(⁂) PROBLEM 3.33 *Skewed Intersection Sight Distance Case C1*

Passenger car A is attempting to cross the below 30 *degrees* skewed intersection to the other side. The minor road with the 45 *mph* design speed has a 5% downgrade in the crossing direction and is controlled with a yield sign. The major road is a 24 *ft* wide two-way highway (one lane in each direction) with a design speed of 65 *mph*.

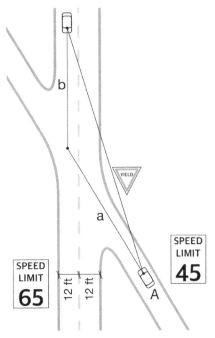

Given the above information, and for a crossing maneuver attempt by the passenger car A, Approach Distance $a =$ _____ ft, and Sight Distance $b =$ _____ ft.

(⁂) Normally you are asked to provide one value.

(⁂) PROBLEM 3.34 *Skewed Intersection Sight Distance Case C2*

Passenger car A attempts a right turn maneuver without stopping using the yield sign posted at the below 30 *degrees* skewed intersection. The minor road has a design speed of 45 *mph* and has no grade. The major road is a 24 *ft* wide two-way highway (one lane in each direction) with a design speed of 65 *mph*.

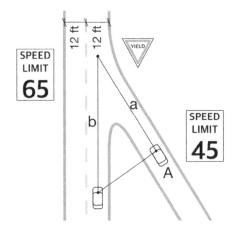

Given the above information, and for a right turn maneuver attempt by the passenger car A, Approach Distance $a =$ _____ ft, and Sight Distance $b =$ _____ ft.

(⁂) Normally you are asked to provide one value.

ROBLEM 3.35 *Intersection Sight Distance Case F*

Truck A is attempting a left turn maneuver from the below major road into a minor road as shown.

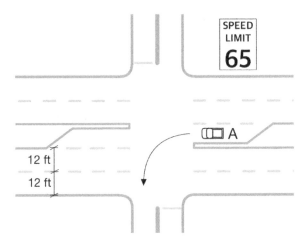

Given that all the above lanes are 12 *ft* wide, a left turn maneuver by truck A requires a sight distance of no less than _____ ft.

PROBLEM 3.36 *Preferable Island Size*
The preferable size for a corner curbed island located at an intersection in an urban area is:

(A) $50\ ft^2$

(B) $75\ ft^2$

(C) $100\ ft^2$

(D) $125\ ft^2$

PROBLEM 3.37 *Deceleration Lane*
The below road is a highway located in a rural area with a design speed of 70 mph. The auxiliary lane shown is part of an unsignalized intersection at a newly constructed area where no information about traffic volume is available yet.

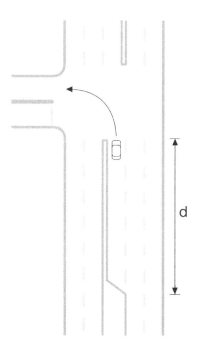

Given the above, the minimum length for this auxiliary lane should most nearly be:

(A) $915\ ft$

(B) $815\ ft$

(C) $875\ ft$

(D) $1,065\ ft$

PROBLEM 3.38 *Storage Length*
The auxiliary lane shown below is part of an unsignalized intersection. The following data can be used:

o Probability of overflow = 0.005
o Critical gap = 6.25 sec
o Follow-up gap = 2.2 sec
o Percentage of trucks = 2%
o Left turn volume = 240 veh/h
o Opposing traffic volume = 1,000 veh/h.

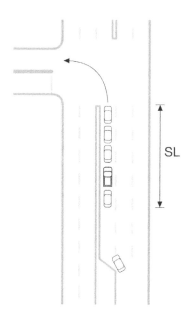

Given the above information, the storage length for this auxiliary lane should be:

(A) $190\ ft - 200\ ft$

(B) $50\ ft - 60\ ft$

(C) $150\ ft - 175\ ft$

(D) $250\ ft - 275\ ft$

PROBLEM 3.39 *Sight Distance for a Track Crossing*
The following figure represents a track crossing with no train-activated warning devices installed. The train runs at a speed of 80 mph while the crossing road design speed is 45 mph.

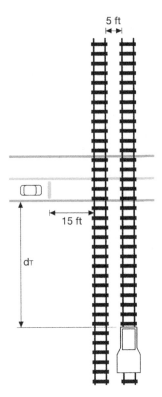

Using AASHTO's recommended design values for speed in first gear, acceleration, dimensions, and perception time; the sight distance d_T for a stopped vehicle is most nearly:

(A) 2,037 ft

(B) 2,170 ft

(C) 2,100 ft

(D) 1,221 ft

PROBLEM 3.40 *Ramp Width*

The desirable traveled way width (*left shoulder + ramp + right shoulder*) for a one-way ramp with moderate traffic volume, 8% trucks, 200 ft inner edge radius, and the ramp should have an option to overtake stalled vehicles, is:

(A) 4 ft + 15 ft + 6 ft

(B) 4 ft + 20 ft + 6 ft

(C) 4 ft + 10 ft + 6 ft

(D) 6 ft + 10 ft + 4 ft

PROBLEM 3.41 *Loop Ramp Design Speed and Superelevation*

The below main highway, which connects with the cloverleaf's loop ramp shown, is located in an urban area and has a design speed of 65 mph.

The available land limits the radius of this ramp to a maximum of 235 ft – i.e., any radius below 235 ft can work.

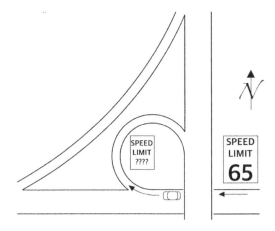

Based on the above, the recommended design speed and superelevation (e) for the loop ramp are most nearly:

(A) 45 mph, e = 6%

(B) 45 mph, e = 8%

(C) 30 mph, e = 4%

(D) 30 mph, e = 6%

PROBLEM 3.42 *A Two-Lane Exit*

An auxiliary lane is provided upstream of a freeway's two-lane exit terminal such that not to reduce the basic number of through lanes while developing capacity for the two-lane exit. The recommended length of this auxiliary lane is:

(A) 750 ft

(B) 1,250 ft

(C) 1,500 ft

(D) 2,000 ft

PROBLEM 3.43 *Highway Taper Exit*
The below highway has a design speed of 70 mph and the speed at the Point of Curve PC on the exit ramp shown below should be 30 mph. The exit taper has a divergence angle of 2° and an upgrade slope of 5% as shown.

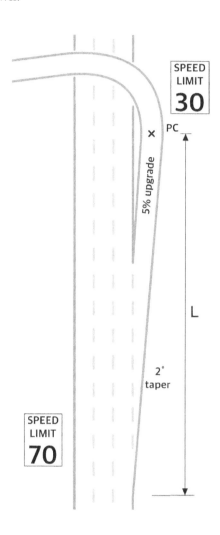

The recommended minimum length from the beginning of the taper to the point of curve for this exit is most nearly:

(A) 760 ft

(B) 690 ft

(C) 420 ft

(D) 520 ft

PROBLEM 3.44 *Highway Taper Entrance*
The below highway has a design speed of 65 mph. The controlled entrance design speed is 45 mph, and the entrance lane width is 16 ft.

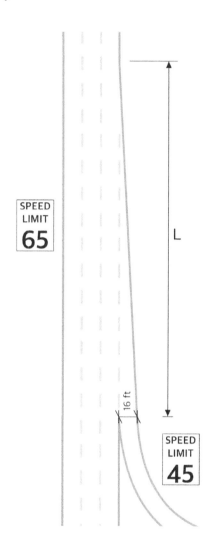

The recommended minimum length from the entrance of the highway to the taper end is most nearly:

(A) 300 ft

(B) 420 ft

(C) 600 ft

(D) 800 ft

PROBLEM 3.45 *Highway Parallel Entrance*

The below highway has a design speed of 65 mph. The average running speed at the ramp's control feature is 44 mph, and the entrance lane width is 12 ft, and there is a downgrade of < 3% at entrance as shown.

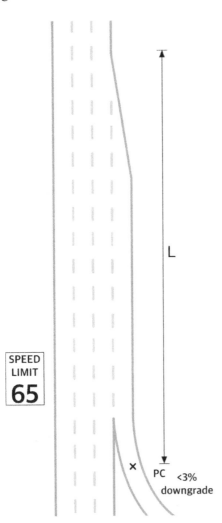

The recommended minimum length from the point of curve PC to the highway end of taper shown is most nearly:

(A) 370 ft

(B) 670 ft

(C) 900 ft

(D) 580 ft

PROBLEM 3.46 *Drainage Channel Cross Section*

The below Drainage channel is designed to collect surface water runoff and drain it to its proper outlet. The clear zones for this section have been set out as shown.

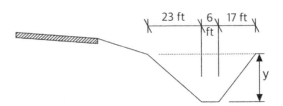

The recommended depth for this channel $'y'$ that produces the most desirable drainage cross section is most nearly:

(A) 4.6 ft

(B) 5.8 ft

(C) 6.9 ft

(D) 8.1 ft

PROBLEM 3.47 *Runout Distance*

In the below road cross section, design speed is 70 mph and ADT is 10,000.

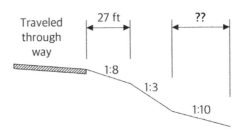

The recommended runout area in the above cross section is most nearly:

(A) 0 ft

(B) 5 ft

(C) 10 ft

(D) 15 ft

PROBLEM 3.48 *Ramp Clear-Zone*

The below loop ramp has the following properties:

- 45 *mph* design speed
- ADT = 4,000
- Radius = 660 *ft*
- Ramp width excluding shoulders = 12 *ft*

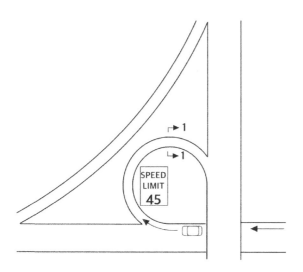

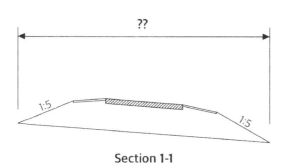

Section 1-1

Based on the clear-zone concept, the recommended total width of the ramp's right of way is:

(A) 65 *ft*

(B) 80 *ft*

(C) 50 *ft*

(D) 55 *ft*

PROBLEM 3.49 *Barrier Offset*

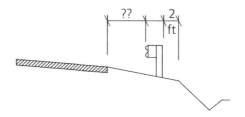

The suggest offset between the above barrier and the edge of the traveled way when the design speed of the through lane is 60 *mph* is most nearly:

(A) 5 *ft*

(B) 6 *ft*

(C) 7 *ft*

(D) 8 *ft*

PROBLEM 3.50 *Length of Barrier*

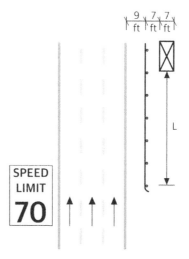

The recommended distance L – length of barrier prior to the obstacle – for the above road with an ADT > 10,000 *veh/day* is:

(A) 220 *ft*

(B) 170 *ft*

(C) 319 *ft*

(D) 75 *ft*

PROBLEM 3.51 *Guardrail at Curb*

The above curb and guardrail are installed in a 40 *mph* road.

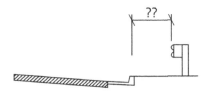

The recommended distance between the curb and the guardrail is:

(A) 5 ft
(B) 6 ft
(C) 7 ft
(D) 8 ft

PROBLEM 3.52 *Slope End Treatments*

Terminating concrete barriers by tapering its end can be used in locations where traffic speed is most nearly:

(A) 45 mph
(B) 40 mph
(C) 55 mph
(D) 65 mph

PROBLEM 3.53 *Sand-Filled Barrels*

The deceleration force generated from a 4,400 *lb* vehicle hitting one 700 *lb* sand filled barrel with a diameter of 3 *ft* at a speed of 90 *mph*, is most nearly:

(A) 32
(B) 46
(C) 23
(D) 64

PROBLEM 3.54 *Self-Restoring Crash Cushions*

Choose from the below list all that applies and that can be identified as self-restoring low maintenance crash cushions:

- Advanced Dynamic Impact Extension Module
- Thrie-Beam Bullnose System
- Compressor
- Narrow Connecticut Impact Attenuation System
- Quad Guard Elite
- Trinity Attenuation Crash Cushion Family
- Smart Cushion Innovations
- Crash Guard

PROBLEM 3.55 *Obstacles Located at Lane Merge Locations*

The below lanes are all 12 *ft* wide, and the taper for the merge is 1: 50.

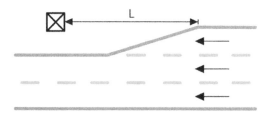

Given the above information, the minimum longitudinal distance identified as L that should be provided for this obstacle is most nearly:

(A) 600 ft
(B) 610 ft
(C) 710 ft
(D) 615 ft

PROBLEM 3.56 *Americans with Disabilities ADA Act for Curbs*

The following are pedestrian facilities' improvements that can be applied to curbs and ramps to bring them into compliance with the ADA act:

I. Cross slopes of adjacent cross slopes should be a minimum of 2.5% but not larger than 3%.
II. A turning space at the top of perpendicular curb ramps of 4 ft by 4 ft
III. The minimum curb ramp slope is 8.33%.
IV. Detectible warning where there is no curb to alert pedestrians with visual needs about the road/curb interface.

(A) II + III + IV
(B) III + IV
(C) I + II + III + IV
(D) II + IV

PROBLEM 3.57 *Retaining Walls*

When retaining walls are required within the clear zone of a freeway, those should be placed at no less than ____ beyond the outer edge of the shoulder:

(A) 2 ft
(B) 4 ft
(C) 6 ft
(D) 8 ft

PROBLEM 3.58 *Traffic Calming Features*

The following can be considered as traffic Calming Features (select all that applies):

- Neighborhood roundabouts
- Traffic Signals
- Road humps
- Law Enforcement
- Diverging Diamond Interchange
- A full cloverleaf interchange

PROBLEM 3.59 *Bridge Walkway Provision for Individuals with Disabilities*

A road's shoulder with pedestrian traffic is connected to an existing bridge. Due to a level difference, the bridge shoulder will be higher than the road's shoulder.

To overcome the level difference between both, taking the ADA act into account, it is allowed to ramp the walkway into the bridge shoulder at a rate of nearly:

(A) 5%
(B) 8.33%
(C) 2%
(D) 4%

PROBLEM 3.60 *Provision for Narrow Walkways*

A 4 ft wide walkway designed for a local road should have a widened area of nearly 5 ft by 5 ft used by pedestrians as a passing area at least every:

(A) 100 ft
(B) 150 ft
(C) 200 ft
(D) 250 ft

SOLUTION 3.1

The requested setback is identified in the *NCEES Handbook version 2.0* as distance M and is measured to the center of the road as identified in the question.

$$M = R - R\cos\left(\frac{\Delta}{2}\right)$$

$$(12 + 0.5 \times 11)\,ft = R \times \left(1 - \cos\left(\frac{48}{2}\right)\right)$$

$$\rightarrow R = 202.4\,ft$$

$$L = \frac{R\,\Delta\,\pi}{180} = \frac{202.4 \times 48 \times \pi}{180} = 169.5\,ft$$

Correct Answer is (D)

SOLUTION 3.2

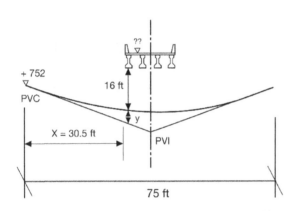

$Y_{PVC} = +752$

$Curve\ elevation = Y_{PVC} + g_1 x + a x^2$

$a = \frac{g_2 - g_1}{2L} = \frac{0.05 - (-0.06)}{2 \times 75} = 7.33 \times 10^{-4}$

$'x'$ is calculated to the edge of the bridge:

$x = \frac{75}{2} - \frac{14}{2} = 30.5\,ft$

Curve Elevation

$= 752 + (-0.06) \times 30.5$
$\qquad\qquad + 7.33 \times 10^{-4} \times 30.5^2$

$= +750.85$

Bridge Deck Elevation

$= 750.85 + 16 + \frac{53}{12} = +771.3$

Correct Answer is (B)

SOLUTION 3.3

In reference to the *NCEES Handbook version 2.0*, Section 5.3.1 Symmetrical Vertical Curve Formula, and the table given in the question – copied here for reference, a vertical curve can be constructed as follows:

Point	Station	Elevation
PVC	0 + 025	72.5
PVI	0 + 275	65
A point on the curve	0 + 150	70

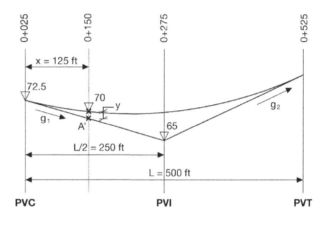

$$g_1 = \frac{65 - 72.5}{275 - 25} = -0.03$$

Determine Elevation of the point on the Back Tangent at station $0 + 150$ located at $x = 125\,ft$.

$A'_{Elevation} = 72.5 - 0.03 \times 125 = 68.75$

$y = 70 - 68.75 = 1.25$

$y = a x^2$

$\rightarrow a = \frac{y}{x^2} = \frac{1.25}{125^2} = 8 \times 10^{-5}$

$a = \frac{g_2 - g_1}{2L}$

$\rightarrow g_2 = 2aL + g_1$

$= 2 \times 8 \times 10^{-5} \times 500 + (-0.03)$

$= 0.05$

Correct Answer is (A)

SOLUTION 3.4
This solution is provided with reference to the Transportation Section Horizontal Design in Chapter 5 of the *NCEES Handbook*.

Given the two curves are identical, the chord C for each curve can be found using trigonometry as follows:

$$C = \sqrt{370^2 + 275^2} = 461\ ft$$

Also, the obstruction is located at the middle ordinate $\rightarrow M = 70\ ft$.

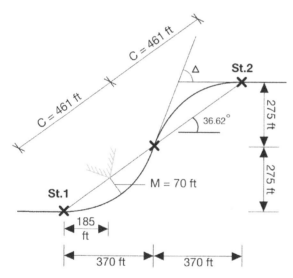

The above two distances C & M can be used to calculate the length of curve L which shall be used to determine Station $St.2$ location as follows:

$$M = R\left(1 - \cos\frac{\Delta}{2}\right)$$

$$C = 2R\sin\frac{\Delta}{2}$$

Divide the above two equation by each other to cancel common coefficients:

$$\frac{M}{C} = \frac{R\left(1 - \cos\frac{\Delta}{2}\right)}{2R\sin\frac{\Delta}{2}}$$

$$\rightarrow \frac{70}{461} = \frac{1 - \cos\frac{\Delta}{2}}{2\sin\frac{\Delta}{2}}$$

$$0.3034\sin\frac{\Delta}{2} = 1 - \cos\frac{\Delta}{2}$$

This equation can be solved with the use of the following identity that can be found in the *NCEES Handbook* Section 1.3 Mathematics:

$$\left(\sin\frac{\Delta}{2}\right)^2 + \left(\cos\frac{\Delta}{2}\right)^2 = 1$$

$$\sin\frac{\Delta}{2} = \sqrt{1 - \left(\cos\frac{\Delta}{2}\right)^2}$$

Substituting this equation into the previous one provides the following:

$$0.3034 \times \sqrt{1 - \left(\cos\frac{\Delta}{2}\right)^2} = 1 - \cos\frac{\Delta}{2}$$

Raise the two sides of the above resultant equation to the power of '2' to get rid of the square root and then rearrange into a quadratic equation:

$$\left(0.3034 \times \sqrt{1 - \left(\cos\frac{\Delta}{2}\right)^2}\right)^2 = \left(1 - \cos\frac{\Delta}{2}\right)^2$$

$$\left(\cos\frac{\Delta}{2}\right)^2 - 1.831\left(\cos\frac{\Delta}{2}\right) + 0.831 = 0$$

The above is a quadratic equation with $a = 1$, $b = -1.831$ and $c = 0.831$ and can be solved as follows:

$$root = \frac{-b \mp \sqrt{b^2 - 4ac}}{2a}$$

$$\cos\frac{\Delta}{2} = \frac{+1.831 \mp \sqrt{1.831^2 - 4 \times 1 \times 0.831}}{2 \times 1}$$

$$= (0.831, 1)$$

The root of '0.831' is used in this case as a root of '1' generates a deflection angle of '0'.

$$\frac{\Delta}{2} = \cos^{-}(0.831) = 33.79°$$

$$\Delta = 67.58°$$

The radius of curvature R can be calculated using any of the previously used equations for either M or C:

$$R = \frac{M}{\left(1 - \cos\frac{\Delta}{2}\right)} = \frac{70}{1 - \cos 33.79} \cong 414.5 \, ft$$

Or

$$R = \frac{c}{2 \sin\frac{\Delta}{2}} = \frac{461}{2 \sin 33.79} \cong 414.5 \, ft$$

The length of the two curves, given they are identical, is calculated as follows:

$$2L = 2 \times \frac{R\Delta\pi}{180} = 2 \times \frac{414.5 \times 67.58 \times \pi}{180}$$

$$\cong 978 \, ft$$

Station $St.2$ is therefore calculated as follows:
$$St.2 = '1 + 00' + 978 \, ft$$

$$= 10 + 78$$

Correct Answer is (C)

SOLUTION 3.5

The solution of this question refers to the material provided by the *NCEES Handbook version 2.0* Chapter 5 Transportation, Section 5.2 Horizontal Design, and Section 5.2.5 Method of Designating Directions which helps identify how bearings work.

First, start with determining the direction for each tangent using the bearing information provided in as follows:

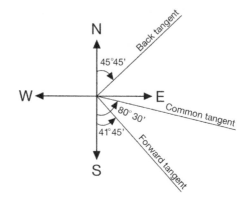

The above tangents are then carried over to construct the following diagram:

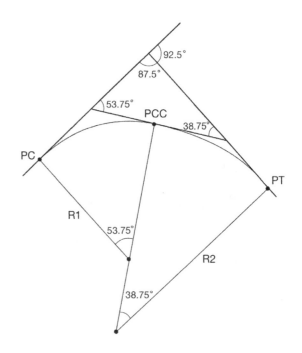

Equations from the basic horizontal curve can be used now to determine the radius for each curve R_1 and R_2 relying on cord length L_1 and L_2 as follows:

$$L = \frac{R \Delta \pi}{180}$$

$$R = \frac{180 L}{\Delta \pi}$$

Given that $L_1 = L_2 = 600\ ft$:

$$R_1 = \frac{180 \times 600}{53.75 \times \pi} = 639.6\ ft$$

$$R_2 = \frac{180 \times 600}{38.75 \times \pi} = 887.2\ ft$$

Ratio $R_1 : R_2 = 1.39 : 1$

Correct Answer is (D)

SOLUTION 3.6
The solution of this question refers to the material provided by the *NCEES Handbook version 2.0* Chapter 5 Transportation, Section 5.2.2 Layout of Two-Centered Compound Curves.

Start with computing all deflection angles in the given diagram as follows:

$$\Delta_1 = 51.45^o$$

$$\Delta_2 = 39.75^o$$

$$\Delta = 51.45^o + 39.75^o = 91.2^o$$

Using the above information, the diagram can be redrawn as follows:

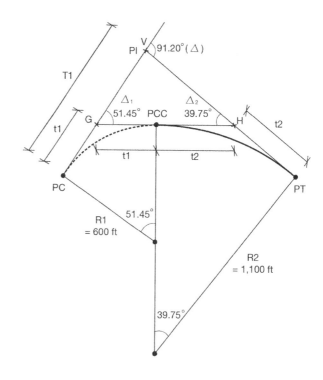

$$t_1 = R_1 \tan\left(\frac{\Delta_1}{2}\right)$$
$$= 600 \times \tan\left(\frac{51.45^o}{2}\right)$$
$$= 289.1\ ft$$

$$t_2 = R_2 \tan\left(\frac{\Delta_2}{2}\right)$$
$$= 1{,}100 \times \tan\left(\frac{39.75^o}{2}\right)$$
$$= 397.7\ ft$$

Using the law of sines:

$$\frac{\overline{VG}}{\sin(\Delta_2)} = \frac{t_1 + t_2}{\sin(\Delta)}$$

$$\frac{\overline{VG}}{\sin(39.75^o)} = \frac{289.1 + 397.7}{\sin(91.20^o)}$$

$$\overline{VG} = 439.3\ ft$$

$$T_1 = t_1 + \overline{VG} = 728.4\ ft$$

Calculate the length of cord L_1:

$$L_1 = \frac{R_1 \Delta_1 \pi}{180}$$

$$= \frac{600 \times 51.45° \times \pi}{180}$$

$$= 538.8 \, ft$$

$$St. @ PCC = PI - T_1 + L_1$$

$$= [3 + 00] - 728.4 \, ft + 538.8 \, ft$$

$$= 1 + 10.4$$

Correct Answer is (D)

SOLUTION 3.7
Chapter 3 Elements of Design, Section 3.3.8.6 Methods of Attaining Superelevation, of AASHTO's *Green Book 7th edition* is referred to in this solution.

The diagram in the following page is redrawn based on Figure 3-8A of the Green Book where rotation of the cross-section is about the centerline of the road section as given in the question.

The Tangent Runout in this case represents the length of the road segment needed to change the outside lane of the Normal Crown cross-section from 2% to a zero superelevation – i.e., transition from cross-section A to cross-section B in the below diagram – and is calculated using Equation 3-30 of the same reference as follows:

$$L_t = \frac{e_{NC}}{e_d} L_s$$

Where L_t is the tangent runout and L_s is the length of the spiral curve (which equals to the length of runoff as given in the question and recommended by AASHTO) – i.e., 250 ft:

$$L_t = \frac{0.02}{0.08} \times 250 = 62.5 \, ft$$

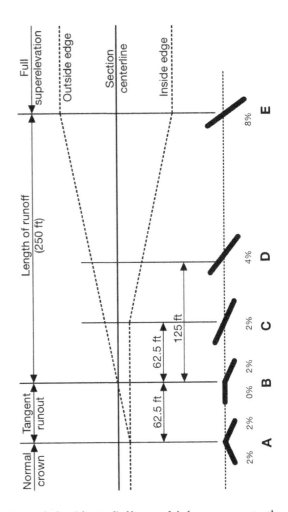

A straight (dotted) line, which represents the "outside edge", is extended from section B where $e = 0\%$ to section E where $e = 8\%$. A similar method is carried out for the inner edge with the dotted line extended from section C at y_{c2} to section E. Elevations for the inside and outside edge are determined using similarity of triangles as follows:

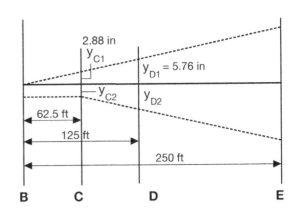

At section C, value y_{C1} represents the elevation of the outer lane's edge taking the cross-section's centerline as datum, and is calculated as follows:

$$y_{C1} = 12\,ft \times \frac{12\,in}{1\,ft} \times 2\% = 2.88\,in$$

y_{C2} is calculated using a similar method and is equal to $2.88\,in$

Cross-section D located midway through the spiral can be computed as follows – where both y_{D1} and y_{D2} are calculated from the centerline of the section:

$$\frac{y_{D1}}{125\,ft} = \frac{2.88\,in}{62.5\,ft}$$

$$y_{D1} = y_{D2} = 5.76\,in$$

The difference between the bottom edge and the top edge of the $24\,ft$ wide cross-section is therefore equals to $5.76 + 5.76 = 11.52\,in$.

Correct Answer is (D)

SOLUTION 3.8
Chapter 3 Elements of Design, Section 3.3.2.2 Side Friction Factor, of AASHTO's *Green Book 7th edition* is referred to in this solution.

The side friction factor is needed here. Since it was not provided in the question, it can be determined from Figure 3-4 of the Green Book as '0.14' for a design speed of $50\,mph$.

Using the above information, Equation 3-7 calculates the requested superelevation:

$$f = \frac{V^2}{15R} - 0.01e$$

$$e = 100 \times \left(\frac{V^2}{15R} - f\right)$$

$$= 100 \times \left(\frac{(50\,mph)^2}{15 \times 900\,ft} - 0.14\right)$$

$$= 4.5\%$$

$$raise = 4.5\% \times 24\,ft = 1.08\,ft\,(= 13\,in)$$

Correct Answer is (A)

SOLUTION 3.9
Chapter 3 Elements of Design, Section 3.3.8.2.1, 2 & 3, of AASHTO's *Green Book 7th edition* are referred to in this solution.

Given that rotation of the road cross-section is about its centerline, Equation 3-23 should be used to calculate the length of runoff L_r:

$$L_r = \frac{(wn_1)\,e_d}{\Delta}\,(b_w) \quad (*)$$

w is the width of one lane ($= 12\,ft$), n_1 is the number of lanes rotated which is one in this case, b_w is the adjustment factor taken as '1' per Table 3-15 and Δ is 0.5%.

$$L_r = \frac{(12 \times 1)\,8\%}{0.5\%} \quad (1)$$

$$= 192\,ft$$

Use Equation 3-25 to compute the proportion of the maximum superelevation attained at PC ($p_{tangent}$) – also observe that v (small letter) in this equation is in ft/sec:

$$\frac{e}{100} < \frac{2.15}{1 + p_{tangent}} \times \frac{v^2}{gR}$$

$$p_{tangent} < 2.15 \times \frac{100}{e} \times \frac{v^2}{gR} - 1$$

$$< 2.15 \times \frac{100}{8} \times \frac{(50 \times 1.47\,ft/sec)^2}{32.2\,ft/sec^2 \times 2{,}500\,ft} - 1$$

$$< 0.8$$

The above reads that superelevation at PC should start at anything lesser than $0.8 \times 8\% = 6.4\%$. The runoff can even start at PC (i.e., $e = 0\%$ at PC and runoff for $L_r = 192\ ft$ to attain an $e = 8\%$ within the curve).

However, if e starts at 8% at PC, there could be a risk of skidding and rollover on the approach tangent. This is because the radius of this curve is above the minimum radius recommended per Tables 3-8 to Table 3-12. Hence why Equation 3-25 recommends starting the runoff on the tangent approach to attain a superelevation of 6.4% or less at PC to provide the necessary safety margins for drivers who are about to enter the curve.

Starting at a superelevation of $e = 0\%$, with a gradient of 0.5%, the runoff required to achieve an $e = 6.4\%$ (a raise above centerline of $12\ ft \times 6.4\% = 0.768\ ft$) or less is calculated as follows:

$$L_{r(0 \to 6.4\%)} = \frac{0.768}{0.5\%} = 153.6\ ft$$

Which represents 80% of the full L_r of $192\ ft$.

Correct Answer is (D)

(*) It is to be noted that Table 3-16a of the same section, that can also be used to determine the length of runoff L_r, is ONLY used when rotation is about the edge of the pavement not its centerline.

SOLUTION 3.10
Chapter 3 Elements of Design, Sections 3.3.8.2 and 3.3.8.6 of AASHTO's *Green Book 7th edition* are referred to in this solution.

Determine the length of runoff L_r using Table 3-16a which is used for cross-sections that rotate about their inner edges for $12\ ft$ wide roads:

$$L_{r@n1=1\ \&\ V=50} = 144\ ft$$

The runout is then calculated in order to understand the transition journey of this cross-section as it approaches the PC – with the use of Equation 3-24:

$$L_t = \frac{e_{NC}}{e_d} L_r$$

$$L_t = \frac{2\%}{6\%} \times 144 = 48\ ft$$

Based on this, the following diagram is drawn based on Figure 3-8B of the Green Book where rotation of the cross-section is about the road's cross-section inner edge:

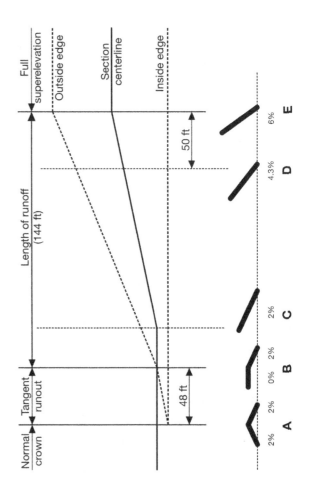

The Tangent Runout L_t is not needed to solve this question, it was calculated here for completeness only. As for the dotted line that extends from A to B (the outer edge of the pavement profile), this line, and for clarity, does not fall on the same path of the outer edge's dotted line from B to E.

The rest of the question can be solved using similarity of triangles as depicted in the below diagram:

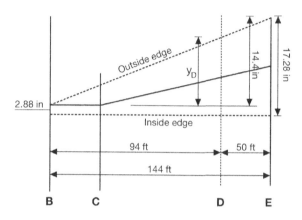

The full raise of the outer edge of the pavement at E is determines by $0.06 \times 24\ ft$, then reduced by $2.88\ in$ to achieve the above triangular shape.

$$\frac{y_D}{94\ ft} = \frac{14.4\ in}{144\ ft}$$

$$y_D = 9.4\ in$$

The raise of the outer edge from the inner edge is therefore calculated as follows:

$Raise = 9.4\ in + 2.88 in$

$\qquad = 12.28\ in\ (e = 4.3\%)$

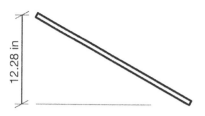

Correct Answer is (D)

SOLUTION 3.11
Chapter 5 of the *NCEES Handbook version 2.0* is referred to, Section 5.2.4 Layout of Reverse Curves Between Parallel Tangents.

Start with sketching the two parallel lines as shown in the below diagram, then add the two connecting reverse curves. The curve on the left is assigned with the radius of $75\ ft$.

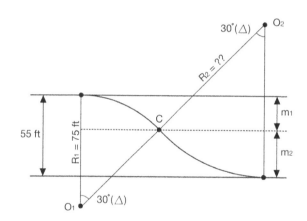

Using trigonometry, and starting from the curve to the right, the following can be established:

$$\cos(30) = \frac{R_1 - m_1}{R_1}$$

$$0.866 = \frac{75\ ft - m_1}{75\ ft}$$

$$m_1 = 10\ ft$$

$$\rightarrow m_2 = 45\ ft$$

Take the same concept to the second triangle as follows:

$$cos(30) = \frac{R_2 - m_2}{R_2}$$

$$0.866 = \frac{R_2 - 45\ ft}{R_2}$$

$$R_2 = 335.8\ ft$$

Compute the arc length of the two curves as follows:

$$L = \frac{R\ \Delta\ \pi}{180}$$

$$L_1 = \frac{75 \times 30 \times \pi}{180} = 39.3\ ft$$

$$L_2 = \frac{335.8 \times 30 \times \pi}{180} = 175.8\ ft$$

$$L_{total} = 39.3 + 175.8\ ft = 215.1\ ft$$

Correct Answer is (B)

SOLUTION 3.12

Chapter 3 Elements of Design, Sections 3.3.9 and 3.3.10, of AASHTO's *Green Book* 7th *edition* are referred to in this solution.

Table 3-24a is used which provides the required widening for a WB-62 truck using a two-lane highway:

$$w_{w_n=24\ ft,\ R=1,000\ ft,\ WB-62} = 2.6\ ft$$

The above value should be corrected using Table 3-25 for a WB-92D truck as follows:

$$w_{WB-92D} = 2.6 + 0.4 = 3.0\ ft$$

This value is then corrected to account for a four-lane roadway as follows:

$$w_{WB-92D,\ 4-lane} = 3.0 \times 2 = 6.0\ ft$$

The above value can also be derived using Figure 3-9 to calculate track width U, Figure 3-10 to derive front Over Hang F_A, Equation 3-34 to calculate the extra width allowance Z, and Equation 3-36 to calculate required horizontal curve width W_c. The only missing factor in Equation 3-36 would be the lateral clearance C which should be provided in the question.

Correct Answer is (C)

SOLUTION 3.13

Chapter 3 Elements of Design, Sections 3.2.2.3 and 3.3.12, of AASHTO's *Green Book* 7th *edition* are referred to in this solution.

The Stopping Sight Distance SSD for the proposed new speed of 50 mph is calculated using Table 3.2 for a 6% downgrade as 474 ft, or by using Equations 3-2 and 3-3 as follows:

$$SSD = 1.47Vt + \frac{V^2}{30\left[\left(\frac{a}{32.2}\right) \mp G\right]}$$

Where deceleration rate a can be taken as recommended in this section as 11.2 ft/sec^2 and the brake reaction time $t = 2.5\ sec$ also can be taken as recommended in this section:

$$SSD = 1.47 \times 50 \times 2.5 + \frac{50^2}{30 \times \left[\left(\frac{11.2}{32.2}\right) - 0.06\right]}$$

$$\cong 474\ ft$$

Equation 3-37 is used to calculate the required HSO from the centerline of the inside lane as follows:

$$HSO = R\left[1 - cos\left(\frac{28.65S}{R}\right)\right]$$

$$= 900 \times \left[1 - cos\left(\frac{28.65 \times 474\ ft}{900\ ft}\right)\right]$$

$$= 31\,ft$$

Ignoring the given original speed, the new object location should be situated $31\,ft$ from the centerline of the one $12\,ft$ lane. Hence:

$$HSO_{edge\ of\ ditch} = 31\,ft - \frac{12\,ft}{2} - 4\,ft - 4\,ft$$

$$= 17\,ft$$

Given the above new distance, the object should move $= 17 - 5 = \mathbf{12\,ft}$ away from its original location in the direction away from the ramp.

Correct Answer is (C)

SOLUTION 3.14

Chapter 5 of the *NCEES Handbook version 2.0*, Section 5.2.1 Basic Curve Elements, is referred to.

Start by determining if station $9 + 35$ falls on the tangent or on the arc as each case requires a different approach.

The steps for calculating arc length L are as follows:

$$C = \sqrt{(3{,}450 - 3{,}330)^2 + (3{,}250 - 2{,}300)^2}$$

$$= 957.5\,ft$$

$$R = \frac{C}{2\sin\left(\frac{\Delta}{2}\right)}$$

$$= \frac{957.5}{2\sin\left(\frac{37.5^o}{2}\right)}$$

$$= 1{,}489.5\,ft$$

$$L = \frac{R\Delta\pi}{180}$$

$$= \frac{1{,}489.5 \times 37.5 \times \pi}{180}$$

$$= 974.9\,ft$$

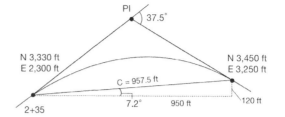

Determine the angle of chord C with a horizontal East/West line – later used to determine other angles and ultimately the requested coordinates for station $9 + 35$:

$$Arctan\left(\frac{120}{950}\right) = 7.2^o$$

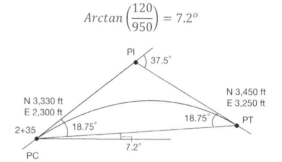

The distance between the requested station $9 + 35$ and station $2 + 35$ is $\ell_1 = 700\,ft$, which means that the requested station falls on the arc as depicted in the following figure that carves out a new part of the circle with a new deflection angle denoted as $\delta_1 = 26.9^o$ – calculations shall follow:

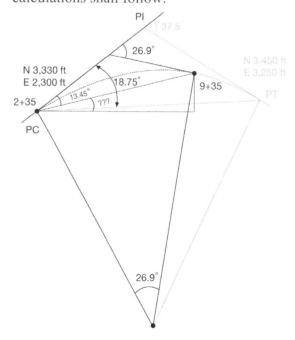

The deflection angle δ_1 the for newly formed the part arc is computed as follows:

$$\frac{\delta_1}{\Delta} = \frac{\ell_1}{L}$$

$$\frac{\delta_1}{37.5°} = \frac{700\ ft}{974.9\ ft}$$

$$\rightarrow \delta_1 = 26.9°$$

In order to determine the coordinates at the requested station $9+35$, a right triangle is constructed as shown. Its hypotenuse (which represents the new Chord C_1) and its angle with a horizontal East/West line shall help determine the requested coordinates.

$$C_1 = 2R\ sin\left(\frac{\delta_1}{2}\right)$$

$$= 2 \times 1,489.5 \times sin\left(\frac{26.9°}{2}\right)$$

$$= 692.9\ ft$$

Chord C_1 angle with the horizontal line is calculated as follows:

$$= 18.75° - 13.45° + 7.2° = 12.5°$$

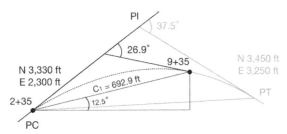

Based on the above, the coordinates for station $9+35$ can now be calculated as follows:

$$Northing = 3,330 + 692.9\ sin(12.5°)$$

$$= 3,480$$

$$Easting = 2,300 + 692.9\ cos(12.5°)$$

$$= 2,976.5$$

Correct Answer is (A)

SOLUTION 3.15
Chapter 5 of the *NCEES Handbook version 2.0*, Section 5.2.1 Basic Curve Elements, is referred to.

Start by determining if station $11 + 50$ falls on the tangent or on the arc as each case requires a different approach.

The steps for calculating arc length L are as follows:

$$C = \sqrt{(1,565 - 1,375)^2 + (3,425 - 2,575)^2}$$

$$= 871\ ft$$

$$R = \frac{C}{2\ sin\left(\frac{\Delta}{2}\right)}$$

$$= \frac{871}{2\ sin\left(\frac{35°}{2}\right)}$$

$$= 1,448.2\ ft$$

$$L = \frac{R\Delta\pi}{180}$$

$$= \frac{1,448.2 \times 35 \times \pi}{180}$$

$$= 884.7\ ft$$

Also determine the angle of chord C with the horizontal East/West line – later used to determine other angles and ultimately the requested coordinates for station $11 + 50$:

$$Arctan\left(\frac{190}{850}\right) = 12.6°$$

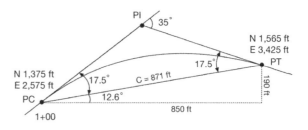

The distance between the requested station $11 + 50$ and station $1 + 00$ is $\ell_1 = 1,050\ ft$, which means that the requested station falls on the tangent beyond the Point of Tangent PT as depicted in the following figure:

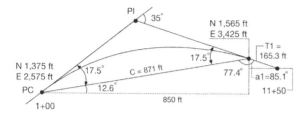

The distance traveled beyond PT to arrive to the requested station is:

$$T_1 = 1{,}050 - 884.7 = 165.3\ ft$$

Tangent T_1 is a hypotenuse and its angle with the vertical line a_1 shall help determine the requested coordinates.

Angle a_1 is determined using triangulation as follows:

$$a_1 = 180° - 77.4° - 17.5° = 85.1°$$

Based on the above, the coordinates for station $11 + 50$ can now be calculated as follows:

$$Northing = 1{,}565 - 165.3\ cos(85.1°)$$
$$= 1{,}550.9$$

$$Easting = 3{,}425 + 165.3\ sin(85.1°)$$
$$= 3{,}589.7$$

Correct Answer is (A)

SOLUTION 3.16
Chapter 3 Elements of Design, Sections 3.3.8.4.3, 3.3.8.4.4 and 3.3.8.4.5 of AASHTO's *Green Book* 7th edition are referred to in this solution.

Use Equations 3-27 and 3-28 (also 3-26) to determine the minimum length for a spiral curve.

Below is Equation 3-27 applied:

$$L_{s,min} = \sqrt{24(p_{min})R}$$

Where p_{min} is the minimum lateral offset between the tangent of the spiral curve and the circular curve – i.e., installing a spiral curve widens the base of the horizontal curve. Unless this distance is given, the Green Book recommends the use of $0.66\ ft$ for minimum calculations.

$$L_{s,min} = \sqrt{24 \times 0.66 \times 900} = 119.4\ ft$$

Equation 3-28 (or Equation 3-26) is used alongside Equation 3-27 as follows:

$$L_{s,min} = 3.15 \frac{V^3}{RC}$$

Where C is the minimum rate of change in lateral acceleration.

The Green Book recommends in this case a value of $4\ ft/sec^3$ which is a value that mimics the minimum experience of drivers' comfort.

$$L_{s,min} = 3.15 \times \frac{65^3}{900 \times 4} = 240.3\ ft$$

Finally, check the desirable values for spiral curve lengths provided in Table 3-19 for a design speed of $65\ mph$:

$$L_{s,Table\ 3-19} = 191\ ft$$

Per the comment provided at the end of Section 3.3.8.4.5 which states that:

"if the desirable spiral curve length shown in Table 3-19 is less than the

minimum spiral curve length determined from Equation 3-26 [similar to Equation 3-28] and Equation 3-27, the minimum spiral curve length should be used in design."

In which case:

$$L_{s,desirable} = min(119.4\ ft, 240.3\ ft, 191\ ft)$$

$$= 119.4\ ft\ (*)$$

Correct Answer is (B)

(*) Another way for determining the appropriate, or desirable, length of a spiral curve, when superelevation is provided, is via calculating the length of runoff explained in the Green Book and in few other examples in this book.

SOLUTION 3.17

Chapter 3 Elements of Design, Section 3.4.6.3 Sag Vertical Curves, of AASHTO's *Green Book 7th edition* is referred to in this solution.

Using Table 3-37 from the Green Book gives a K-value with a design speed of 65 *mph* of:

$$K = 157$$

$$A = |g_2 - g_1|$$

$$= |4\% - (-7\%)|$$

$$= 11\%$$

$$L = KA$$

$$= 11 \times 157$$

$$= 1,727\ ft$$

Correct Answer is (A)

SOLUTION 3.18

Chapter 3 Elements of Design, Sections 3.4.5 Emergency Escape Ramps of AASHTO's *Green Book 7th edition* is referred to in this solution.

Using Table 3-34 determines that the Rolling Resistance for Pea Gravel is 250 *lb*/1,000 *lb GVW*.

Use Equation 3-40 to predict if the truck will stop at the first ramp, if not, Equation 3-41 will be used to determine the truck speed after exiting the first ramp onto the second ramp.

$$L = \frac{V^2}{30(R \mp G)}$$

$$= \frac{100^2}{30(0.25 + 0.1)}$$

$$= 952.4\ ft$$

Given this distance is larger than the first arrestor bed which is 500 *ft* long, the speed of the truck exiting the first arrestor bed V_f will be calculated, after that, the initial calculation will be repeated as follows:

$$V_f^2 = V_i^2 - 30L(R \mp G)$$

$$V_f = \sqrt{V_i^2 - 30L(R \mp G)}$$

$$= \sqrt{100^2 - 30 \times 500 \times (0.25 + 0.1)}$$

$$= 68.9\ mph$$

$$L = \frac{68.9^2}{30(0.25 + 0.05)} = 527.8\ ft$$

$$L_{required} = 500 + 527.8 = 1,027.8\ ft$$

Correct Answer is (A)

SOLUTION 3.19

NCEES Handbook version 2.0 Chapter 5 Transportation and the vertical curve equations are referred to in this solution.

The parabolic equation for curve elevation defined in Section 5.3.1 is used and it follows the y and x axis/datum defined in the figure below:

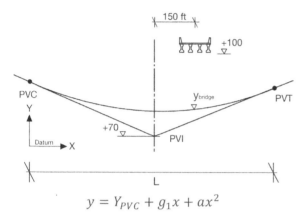

$$y = Y_{PVC} + g_1 x + ax^2$$

The constants in the above equation are defined and calculated as follows:

y is the curve elevation taking datum as defined in the previous figure – the location of datum can change up or down, the final result will not change though.

Curve elevation at the point of interest y_{bridge} is calculated using the bottom of bridge level (+100) and the clearance 16 ft:

$$y_{bridge} = 100 - 16 = 84$$

x is the distance from PVC to the point of interest, in which case, given the curve is symmetrical:

$$x = L/2 + 150\ ft$$

Y_{PVC} is the elevation of PVC, and when no station information is given, it can be calculated using the back tangent and the PVI elevation (+70) as follows:

$$Y_{PVC} = 70 + 0.05 \times L/2$$

Finally, a is the parabola constant and is calculated as follows:

$$a = \frac{g_2 - g_1}{2L} = \frac{0.09}{2L}$$

Substituting the above in the parabolic equation gives us the following:

$$y_{bridge} = Y_{PVC} + g_1 x + ax^2$$

$$84 = \left(70 + 0.05 \times \frac{L}{2}\right)$$
$$-0.05\left(\frac{L}{2} + 150\right)$$
$$+ \frac{0.09}{2L} \times \left(\frac{L}{2} + 150\right)^2$$

The above is a quadratic equation that can be reduced to the following equation:

$$L^2 - 1{,}311.11 L + 90{,}000 = 0$$

This quadratic equation has the following constants: $a = 1$, $b = -1{,}311.11$ and $c = 90{,}000$ and can be solved as follows:

$$root = \frac{-b \mp \sqrt{b^2 - 4ac}}{2a}$$

$$L = \frac{+1{,}311.1 \mp \sqrt{(-1{,}311.1)^2 - 4 \times 1 \times (90{,}000)}}{2 \times 1}$$

$$= (72.67\ ft, 1{,}238.4\ ft)$$

A length of $1{,}238.4\ ft$ is more realistic compared to a length of $72.67\ ft$ in this case.

$$L = 1{,}238.4\ ft$$

Correct Answer is (B)

SOLUTION 3.20

Chapter 3 Elements of Design, Section 3.4.6.4 Sight Distance at Undercrossings, of AASHTO's *Green Book* 7th edition is referred to in this solution.

Sight distance is calculated using Section 3.2.2 as follows:

$$S = SSD = 1.47Vt + 1.075\frac{V^2}{a}$$

$$S = 1.47 \times 65 \times 2.5 + 1.075 \times \frac{65^2}{11.2}$$

$$= 644.4 \, ft$$

Table 3.1 (*) can be used for this purpose as well and it gives a designed SSD of 645 ft.

Also, the algebraic difference between grades $A = 0.04 - (-0.05) = 0.09$ will be used in the coming equations.

Use Equation 3-55 for $S > L$ – which is a condition that shall be verified later:

$$L = 2S - \frac{800(C-5)}{A}$$

$$= 2 \times 644.4 - \frac{800(16-5)}{9}$$

$$= 311 \, ft < 644.4 \, ft \rightarrow ok$$

Use Equation 3-56 for $S < L$ – which is a condition that shall be verified later:

$$L = \frac{AS^2}{800(C-5)}$$

$$= \frac{9 \times 65^2}{800(16-5)}$$

$$= 424.7 \, ft < 644.4 \, ft \rightarrow not \, ok$$

Based on the above $L = 311 \, ft$

Correct Answer is (D)

(*) It is to be noted that the flat/level SSD per Table 3-1 and Equation 3-2 has been used although the under-crossing would fall on a slope.

An extra effort could be made to determine the upgrade at the exact location underneath the crossing using the parabolic equation of the vertical curve – the exact location of the bridge should be known in this case, and SSD can be calculated using the below equation – or Table 3.2 of the Green Book instead – where G would be an upgrade in this case given the bridge location:

$$SSD = 1.47Vt + \frac{V^2}{30\left[\left(\frac{a}{32.2}\right) \mp G\right]}$$

The upgrade G however would be much less than 4% which is the slope of the front tangent.

Taking this information into Table 3-2, along with a design speed of 65 mph, gives us an SSD value of anything between 612 ft to 645 ft.

Take this information back to Equation 3-55:

$$L = 2S - \frac{800(C-5)}{A}$$

$$= 2 \times (612 \, to \, 645) - \frac{800(16-5)}{9}$$

$$= 246 \, ft \, to \, 312 \, ft$$

The above information does not significantly change the outcome of the solution, and given the options in the question, the correct answer remains as D.

SOLUTION 3.21

The *NCEES Handbook version 2.0* Chapter 5 Transportation and the vertical curve equations are referred to in this solution.

Start with constructing the curve's parabolic equation as defined in Section 5.3.1 as follows:

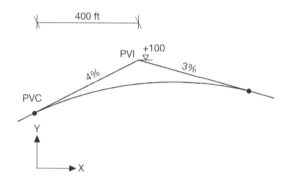

$$y = Y_{PVC} + g_1 x + ax^2$$

$$Y_{PVC} = 100 - 0.04 \times 400 = 84$$

$$a = \frac{g_2 - g_1}{2L} = \frac{-0.07}{1600} = -4.375 \times 10^{-5}$$

Substituting the above back into the parabolic equation:

$$y = 84 + 0.04x + (-4.375 \times 10^{-5})x^2$$

Gradient of the curved slope is represented by the first derivative of the above equation which is as follows:

$$\frac{dy}{dx} = 0.04 + (-8.75 \times 10^{-5})x$$

$$-0.015 = 0.04 - (8.75 \times 10^{-5})x$$

$$x = 628.6 \, ft$$

$$\text{Station}_{\frac{dy}{dx} = -1.5\%} = [1 + 00] + (628.6 \, ft)$$

$$= 7 + 28.6$$

Correct Answer is (C)

SOLUTION 3.22

Chapter 3 Elements of Design, Section 3.4.6.2 Crest Vertical Curves, of AASHTO's *Green Book 7th edition* is referred to in this solution.

Use Equation 3.44 assuming $S < L$ (to be verified later), and $A = 3 + 3 = 6$, sight distance can be calculated as follows:

$$S = \sqrt{\frac{2{,}158 \, L}{A}}$$

$$= \sqrt{\frac{2{,}158 \times 1{,}000}{6}}$$

$$= 599.7 \, ft < L \quad \text{ok}$$

Check Equation 3.45 with $S > L$ for further confirmation:

$$S = \frac{1}{2}\left(L + \frac{2{,}158}{A}\right)$$

$$= \frac{1}{2}\left(1{,}000 + \frac{2{,}158}{6}\right)$$

$$= 679.8 \, ft < L \quad \text{Not ok, use Eq. 3.44}$$

Sight distance is calculated using Section 3.2.2 as follows:

$$S = SSD = 1.47Vt + 1.075 \frac{V^2}{a}$$

$$599.7 = 1.47 \times 2.5V + 1.075 \frac{V^2}{11.2}$$

$$0.096V^2 + 3.675V - 599.7 = 0$$

This quadratic equation has the following constants: $a = 0.096$, $b = 3.675$ and $c = -599.7$ and can be solved as follows:

$$\text{root} = \frac{-b \mp \sqrt{b^2 - 4ac}}{2a}$$

$$V = \frac{-3.675 \mp \sqrt{(3.675)^2 - 4 \times 0.096 \times (-599.7)}}{2 \times 0.096}$$

$$= (62.2 \, mph, -100 \, mph)$$

A design speed of 62.2 mph is more realistic compared to a negative speed.

$$V = 62.2\ mph$$

Alternative and quicker method

With the use of Figure 3-36, substituting the curve length of $1,000\ ft$ at its x axis and A of $'6'$ at its y axis, the required speed can be interpolated as $62.2\ mph$.

Correct Answer is (B)

SOLUTION 3.23

Chapter 3 Elements of Design, Section 3.4.6.2.2 Design Controls: Passing Sight Distance, of AASHTO's *Green Book* 7th edition is used in this solution.

Using Table 3-36 (for crest curves) from the Green Book gives us a K-value with a design speed of 60 mph of:

$$K = 357$$

$$A = |g_2 - g_1|$$
$$= |-3\% - (+3\%)|$$
$$= 6\%$$

$$L = KA$$
$$= 6 \times 357$$
$$= 2,142\ ft$$

Correct Answer is (A)

SOLUTION 3.24

There are two methods that can be used to solve this question: a short method that requires memorizing two equations, and a longer one which requires deriving an equation and solving it quadratically. The two methods are presented below.

The short method:
In this method we need to establish three points as shown in the below graph: E (the point of interest), F and G.

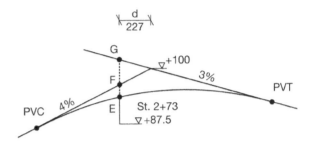

Knowing their elevations, the correction factor S can be determined, and the curve length L can then be calculated as follows:

$$S = \sqrt{\frac{Elev.E - Elev.G}{Elev.E - Elev.F}}$$

$$L = \frac{2d(S+1)}{(S-1)}$$

$d = Station\ @\ PVI - Station\ @\ E$
$= [5 + 00] - [2 + 73]$
$= 227\ ft$

$Elev.E = +87.5$

$Elev.F = 100.0 - 227 \times 0.04 = +90.92$

$Elev.G = 100.0 + 227 \times 0.03 = +106.81$

$$S = \sqrt{\frac{87.5 - 106.81}{87.5 - 90.92}} = 2.3761$$

$$L = \frac{2 \times 227 \times (2.3761 + 1)}{(2.3761 - 1)} = 1,113.8\ ft$$

The long method:
This method involves constructing the parabolic equation for the curve, similar to some of the previous solutions (check solution 3.19 for instance) and solving it for L.

$$y = Y_{PVC} + g_1 x + a x^2$$

x is the distance from PVC to the point of interest (E), in which case:

$$x = L/2 - 227$$

Y_{PVC} is the elevation of PVC, and when no station information is given, it can be calculated using the back tangent and the PVI elevation (+100) as follows:

$$Y_{PVC} = 100 - 0.04 \times \frac{L}{2}$$

a is the parabola constant and is calculated as follows:

$$a = \frac{g_2 - g_1}{2L} = \frac{-0.035}{L}$$

Substituting the above in the parabolic equation generates the following equation:

$$y_E = Y_{PVC} + g_1 x + a x^2$$

$$87.5 = (100 - 0.02L)$$
$$+ 0.04 \left(\frac{L}{2} - 227\right)$$
$$- \frac{0.035}{L} \times \left(\frac{L}{2} - 227\right)^2$$

The quadratic equation can be reduced to the following equation:

$$L^2 - 1{,}298.86 L + 206{,}116 = 0$$

This quadratic equation has the following constants: $a = 1$, $b = -1{,}298.86$ and $c = 206{,}116$ and can be solved as follows:

$$root = \frac{-b \mp \sqrt{b^2 - 4ac}}{2a}$$

$$L = \frac{+1{,}298.86 \mp \sqrt{(-1{,}298.86)^2 - 4 \times 1 \times (206{,}116)}}{2 \times 1}$$

$$= (185.1 \, ft, 1{,}113.8 \, ft)$$

A length of $1{,}113.8 \, ft$ is more realistic compared to a length of $185.1 \, ft$ in this case.

$$L = 1{,}113.8 \, ft$$

Correct Answer is (C)

SOLUTION 3.25

The *NCEES Handbook version 2.0* Chapter 5 Transportation and the vertical curve equations are referred to in this solution.

Per Section 5.3.1, the lowest (or highest point) is located at x_m which is calculated as follows:

$$x_m = -\left(\frac{g_1}{2a}\right)$$

$$a = \frac{g_2 - g_1}{2L} = \frac{0.03 - (-0.04)}{2(835 - 120)} = 4.9 \times 10^{-5}$$

$$x_m = -\left(\frac{-0.04}{2 \times 4.9 \times 10^{-5}}\right) = 408.6 \, ft$$

Construct the equation of the curve to determine the required elevation at $x = x_m = 408.6 \, ft$ as follows:

$$y = Y_{PVC} + g_1 x + a x^2$$

$Y_{PVC} = 77 + 0.04 \times \frac{835-120}{2} = 91.3$

$y = 91.3 - 0.04x + 4.9 \times 10^{-5}x^2$

$y_{@408.6} = 91.3 - 0.04(408.6)$
$\qquad\qquad + 4.9 \times 10^{-5}(408.6)^2$

$\qquad = +83.14$

Correct Answer is (A)

SOLUTION 3.26
In reference to Section 3.5 of AASHTO's Green book – Combinations of Horizontal and Vertical Alignment – along with design general knowledge, the following conclusions can be made per statement.

I. Horizontal curves should not begin close to the top of a crest curve.

A horizontal curve should not begin or end near the top of a crest vertical curve. Drivers in this case cannot recognize where the curve begins or where it ends, especially when visibility is in question such as during nighttime. For safety purposes, horizontal curves should lead vertical curves and they should be made longer and more visible.

This makes Statement I true.

II. The vertical curve should lead the horizontal curve.

As indicated in the previous point, horizontal curves are the ones that should lead vertical profiles, they should be longer as well, and it is preferred to locate the PC before the PVC whenever possible.

This marks Statement II as incorrect.

III. It is best to locate the flat portion of the vertical profile near the flat cross section of the superelevation transition.

For drainage purposes, the above criteria will not work and will cause water to pond on the street. Vertical and horizontal curves when designed together should avoid flat portions to be aligned with each other.

This marks Statement III as incorrect.

IV. Sharp horizontal curves should not be introduced near the bottom of steep grades approaching a sag curve.

If such case is introduced in the design, the view of the road ahead will be shortened resulting in reduced safety measures. Overall appearance of the road will be distorted as well causing erratic operation especially at nighttime and when visibility is impaired.

This makes Statement IV true.

Correct Answer is (D)

SOLUTION 3.27
Chapter 4 Cross-Section Elements, Section 4.20 On-Street Parking, of AASHTO's *Green Book 7th edition* is used to provide solution to this question.

In reference to page 4-87 of the Green Book, and to Figure 4-25, the parking lane is recommended to end at least $20\ ft$ in advance of the intersection.

Correct Answer is (B)

SOLUTION 3.28

Chapter 5 Local Roads and Streets, Section 5.2.1.5 Grades, of AASHTO's *Green Book* 7th edition is used here.

In reference to page 5-4 of the Green Book, and Table 5-2, the maximum grade that can be used for local roads with design speed of 25 *mph* in a mountainous rural area is 15%.

Correct Answer is (D)

SOLUTION 3.29

Chapter 6 Collector Roads and Streets, Section 6.2.2.1 Width of Roadway, of AASHTO's *Green Book* 7th edition is used here.

In reference to page 6-6 of the Green Book, and Table 6-5 along with comment (a) below this table, the minimum width for roadways with design volume under 250 *veh/day* is 18 *ft*, add to that a 2 *ft* per the table above it to account for the roadway's shoulder, this makes the traveled way width:

$$= 18 + 2 + 2 = 22\ ft.$$

Correct Answer is (C)

SOLUTION 3.30

In reference to page 7-20 of AASHTO's Green Book, start with calculating the width of the median in order to determine which of the three methods or cases of attaining superelevation apply:

$$W_{median} = 70 - 4 \times 12 - 2 \times 4 = 14\ ft$$

Based on this, the best method for attaining superelevation for a divided road with median width $< 15\ ft$ is to deal with the entire traveled way as one plane as explained in option C.

Moreover, and with regard to option D which states that no superelevation is required in urban areas, check page 7-38 of the Green Book which confirms this for low speed curbed arterial streets only. However, the given speed of this arterial is 60 *mph* which is not considered as low speed and hence superelevation is still required in this case.

Correct Answer is (C)

SOLUTION 3.31

Chapter 8 Freeways, Section 8.4.2 Medians, of AASHTO's *Green Book* 7th edition is used here.

The description provided in page 8-13 of the Green Book states that the preferred minimum width for medians located in urban areas for six-lane freeways along with a $DDHV > 250\ veh/h$ is 26 ft.

Correct Answer is (C)

SOLUTION 3.32

Chapter 9 Intersections, Section 9.5.3.2.1 Case B1 and Section 9.5.3.2.2 Case B2, of AASHTO's *Green Book* 7th edition are referred to in this solution.

<u>Left Turn from the Minor Road</u>

For case B1 (identified as $b1$ in this example for ease of reference), use Table 9-6 for passenger cars with $t_g = 7.5\ sec$, add to this an additional 0.5 *sec* for each additional lane (in this case 1.5 lanes considering the median as half lane being 6 ft wide). Also add to it 0.2 *sec* for each percent of a grade as follows:

$t_g = 7.5 + 1.5 \times 0.5 + 5 \times 0.2 = 9.25\ sec$

Moreover, last paragraph of page 9-45 states that there is no requirement to adjust the sight distance for major road upgrades as both the minor and the major roads tend to be at the same grade when departing from the intersection.

$$b1 = ISD = 1.47 \times V_{major} \times t_g$$
$$= 1.47 \times 65 \times 9.25$$
$$= 883.8 \, ft$$

Right Turn from the Minor Road

For case B2 (identified as $b2$ in this example for ease of reference), use Table 9-8 for passenger cars with $t_g = 6.5 \, sec$, add to this an additional $0.1 \, sec$ for each percent of a grade as follows:

$$t_g = 6.5 + 5 \times 0.1 = 7 \, sec$$

$$b2 = ISD = 1.47 \times V_{major} \times t_g$$
$$= 1.47 \times 65 \times 7$$
$$= 668.9 \, ft$$

Correct Answer: $b1 = 884 \, ft$, $b2 = 669 \, ft$

SOLUTION 3.33

Chapter 9 Intersections, Section 9.5.3.3.1 Case C1 and Section 9.5.4 Effect of Skew, of AASHTO's *Green Book* 7th edition are referred to in this solution.

Let's resolve the skew matter first by referencing the relevant Section 9.5.4. Page 9-58 & 59 of this section mention that the sight triangles [for cases B and C] are applicable to oblique-angle intersections. The width w however should be adjusted by dividing it by the sine of the skew angle.

Equation 9-2 is used to calculate b. Also, Table 9-12 along with the adjustment to the downgrade from Table 9-5 are used to calculate t_a as $5.2 \, sec$, which needs to be adjusted to account for downgrade using the relevant adjustment factor from Table 9-5 of '1.1'. This gives us an adjusted value for t_a of:

$$t_a = 5.2 \times 1.1 = 5.75 \, seconds$$

$$t_g = t_a + \frac{w + L_a}{0.88 \, V_{minor}}$$

$$w = \frac{24 \, ft}{\sin 30} = 48 \, ft$$

The length of a passenger vehicle L_a is picked up from Table 2-4a found in page 2-56 of the Green Book as $19 \, ft$.

$$t_g = 5.75 + \frac{48 \, ft + 19 \, ft}{0.88 \times 45} = 7.44 \, sec$$

However, it is important to check the flip notes under Table 9-12, and the paragraph that follows. They stipulate that "the value of t_g should equal or exceed the appropriate travel time for crossing the major road from a stop-controlled approach as shown in Table 9-8". This is done in reference to Table 9-8 as follows:

$$t_g = 6.5 + 0.1 \, sec \times 5 + 2 \times 0.5 = 8 \, sec \, (*)$$

$$b = 1.47 \, V_{major} \, t_g$$
$$= 1.47 \times 65 \times 8$$
$$= 764.4 \, ft$$

Minor road approach a is calculated using Table 9-12 and Table 9-5 to account for the downgrade as follows:

$$a = 275 \times 1.1 + \frac{12 \, ft}{\sin 30} = 326.5 \, ft \, (**)$$

Correct Answer: $a = 327 \, ft$, $b = 764 \, ft$

(*) It is important to observe that the inclined width in this case is equivalent to four lanes (i.e., 48 ft), Table 9-8 is designed for a two-lane width, in which case the flip notes of Table 9-10, by which adding 0.5 $seconds$ per extra lane "for each additional lane in excess of two" were used to calculate the final value for t_g

(**) Table 9-12 provides distance a_1 which is the same distance identified in Figure 9-16 of Section 9.5.2.1 Approach Sight Triangles.

Distance a in this question represents the approach sight distance a_2 shown in Figure 9-16. Page 9-37 states that distance a_2 equals to a_1 added to it the width of the lane(s) departing from the intersection on the major road to the right, in which case, this is equivalent to an inclined 12 ft.

SOLUTION 3.34
Chapter 9 Intersections, Section 9.5.3.3.2 Case C2 and Section 9.5.4 Effect of Skew, of AASHTO's *Green Book 7th edition* are referred to in this solution.

In reference to Solution 3.33, the oblique-angle intersection is found to be applicable to sight distances for cases C and B as well.

Approach a in this question equals to distance a_1 referred to in Figure 9-16 of the Green Book, which is set at 82 ft as discussed in Section 9.5.3.3.2.

$$a = a_1 = 82\ ft$$

Distance b is measured using time gap t_g defined in Table 9-14 which equals to 8 sec for passenger cars with no further adjustments to right turns.

$$t_g = 8\ sec$$

$b = 1.47\ V_{major}\ t_g$

$= 1.47 \times 65 \times 8$

$= 764.4\ ft$

Correct Answer: $a = 82\ ft, b = 764\ ft$

SOLUTION 3.35
Chapter 9 Intersections, Section 9.5.3.6 Case F, of AASHTO's *Green Book 7th edition* is referred to in this solution.

Page 9-56 mentions that the "sight distance along the major road to accommodate left turns is the distance traversed at the design speed of the major road in the travel time for the design vehicle given in Table 9-16".

Using this table, with a truck crossing one lane "in excess of one lane":

$$t_g = 6.5 + 1 \times 0.7 = 7.2\ sec$$

$b = 1.47\ V_{major}\ t_g$

$= 1.47 \times 65 \times 7.2$

$= 688\ ft$

Correct Answer is: 688 ft

SOLUTION 3.36
Chapter 9 Intersections, Section 9.6.3.5 Island Size and Designation, of AASHTO's *Green Book 7th edition* is used here.

Page 9-74 of this section states that the preferable size for curbed corner islands for both urban and rural locations is 100 ft^2.

Correct Answer is (C)

SOLUTION 3.37
Chapter 9 Intersections, Section 9.7.2 Deceleration Lanes, of AASHTO's *Green Book 7th edition* is referred to in this solution.

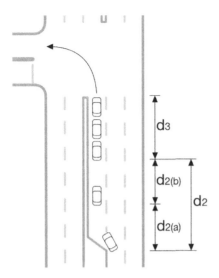

Page 9-94 of this section states that the length of the auxiliary lane is the sum of the three components shown in the above figure and they are: (1) the lane change distance $d_{2(a)}$, (2) the deceleration distance $d_{2(b)}$, and (3) the storage distance d_3.

Distance d_2 can be picked up from Table 9-20 as $815\ ft$ for roads with speed of $70\ mph$.

As for the storage distance d_3 and given this is a newly constructed road with no information on volume of traffic available, page 9-98 recommends the use of a minimum of $100\ ft$ for high speed and rural locations.

$$d = d_2 + d_3 = 815 + 100 = 915\ ft$$

Correct Answer is (A)

SOLUTION 3.38
Chapter 9 Intersections, Section 9.7.2.2 Storage Length, of AASHTO's *Green Book 7th edition* is referred to in this solution.

The solution is provided using two methods: (1) the shorter method using Table 9-22, and (2) the longer method using Equations 9-3 and 9-4.

Although the two methods provide almost similar answers with the tables being slightly more conservative, the assumptions behind both Tables 9-21 or 9-22 may differ in a question during the exam forcing the use of the equations instead.

The shorter method using Table 9-22:
The below list which is given in the question coincide with the assumptions of Table 9-22:

- Probability of overflow = 0.005
- Critical gap = $6.25\ sec$
- Follow-up gap = $2.2\ sec$
- Percentage of trucks = 2%

The percentage of trucks determines the assumed storage length per vehicle which equals to $25\ ft$ per Table 9-23.

Using a left turn volume of $240\ veh/h$ and an opposing volume of $1,000\ veh/h$ in Table 9-22 gives us a storage length of:

$$SL = 275\ ft$$

The longer method using Eq.s 9-3 and 9-4:

$$c = \frac{V_o\, e^{-V_o t_c/3{,}600}}{1 - e^{-V_o t_f/3{,}600}}\ (*)$$

$$= \frac{1{,}000\, e^{-1{,}000 \times 6.25/3{,}600}}{1 - e^{-1{,}000 \times 2.2/3{,}600}}$$

$$= 385\ veh/h$$

$$SL = \left(\frac{\ln[P(n>N)]}{\ln\left[\frac{v}{c}\right]} - 1\right) \times VL$$

$$= \left(\frac{\ln[0.005]}{\ln\left[\frac{240}{385}\right]} - 1\right) \times 25$$

$$= 255\ ft\ (**)$$

Correct Answer is (D)

(*) This equation is corrected in the Green Book's October 2019 errata downloadable through the following link:

https://downloads.transportation.org/gdhs-7-errata.pdf

Also, the same (correct) equation can be found in the *HCM Manual 6th Edition* Equation 20-35 page 20-21.

(**) Table 9-22 provides a more conservative result.

SOLUTION 3.39

Chapter 9 Intersections, Section 9.12.4 Sight Distance [for RailRoad-Highway Grade Crossings], of AASHTO's *Green Book 7th edition* is referred to in this solution.

The described case in this question belongs to case B which represents a departure sight distance. In which case Equation 9-6 should be used. Notice however that value W (set already in the Green Book's relevant Equation as $5\,ft$) shall be adjusted to accommodate the entire width for the two tracks – i.e., $W = 5 + 5 + 5 = 15\,ft$ (*).

$$d_T = AV_T \left[\frac{V_G}{a_1} + \frac{L+2D+W-d_a}{V_G} + J \right]$$

$$= 1.47 \times 80 \times \left[\frac{8.8}{1.47} + \frac{73.5 + 2 \times 15 + 15 - \frac{8.8^2}{2 \times 1.47}}{8.8} + 2 \right]$$

$$= 2{,}171\,ft \quad (**)$$

Correct Answer is (B)

(*) Page 9-168 states that "in order for vehicles to cross two tracks from a stopped position... sight distances along with the railroad should be determined with a proper adjustment for the W value". Because of this, Table 9-29 cannot be used to provide a solution here.

(**) All values which belong to this equation are given in the relevant section of the Green Book.

SOLUTION 3.40

The following sections are referred to from the AASHTO *Green Book 7th edition*:

- Section 3.3.11 Widths of Turning Roadways at Intersections, Table 3-27, page 3-109.
- Section 10.9.6.3.1 Width and Cross Section, page 10-121.
- Section 10.9.6.3.2 Shoulder Widths and Lateral Offset, page 10-121.

The question describes a moderate traffic volume with 8% trucks. Checking the last paragraph of Section 10.9.6.3.1, this belongs to Traffic Condition B.

Table 3-27, page 3-109, for case II (provision for passing stalled vehicles), and condition B (described above), gives us a traveled way width of $20\,ft$ for a $200\,ft$ inner radius ramp. The Ramp width however, per Section 10.9.6.3.2, should not be less than the same condition for Case I – i.e., $15\,ft$ – when shoulders are added.

Section 10.9.6.3.2 (1st bullet point) provides desirable shoulders' width to the ramp's right and left side. It states that the combined shoulders' width should be deducted from the traveled way while maintaining the minimum ramp width stated above (3rd bullet point) – i.e., $15\,ft$. This makes the desirable ramp width when a shoulder width of $4\,ft$ and $6\,ft$ are selected (or deducted) as follows:

$$20\,ft - 4\,ft - 6\,ft = 10\,ft < 15\,ft$$

Hence, the desirable total traveled way width should be:

Left shoulder	Ramp	Right Shoulder
$4\,ft$	$15\,ft$	$6\,ft$

Correct Answer is (A)

SOLUTION 3.41

The following sections are referred to from the AASHTO *Green Book 7th edition*:

- Section 10.9.6.2 General Ramp Design Considerations, page 10-105 Table 10-1, and page 10-106.
- Section 3.3.3.3 Minimum Radius, Table 3-7, page 3-34.

The design speed of this ramp, as presented in Table 10-1 for a 65 mph highway, ranges from 30 mph to 55 mph.

Section 10.9.6.2.4 page 106 states that loop ramps in urban areas with limited land should have the lower speed of this range selected. In which case the recommended speed for this ramp is 30 mph.

The recommended speed of 30 mph along with a land limit of $R < 235\ ft$ are checked against Table 3-7 page 3-34 for the selection of superelevation as follows (*):

Design Speed mph	Maximum (e)	Radius ft
30	6%	231

Correct Answer is (D)

(*) The solution could have referenced Table 3-7 right away without the need for Section 10.9.6.2, and the question options could be checked one by one against values in this table.

SOLUTION 3.42

Section 10.9.6.6.6 Two-Lane Exits of AASHTO *Green Book 7th edition* states that "a distance of approximately 1,500 ft is recommended to develop the full capacity of a two-lane exit."

Correct Answer is (C)

SOLUTION 3.43

Chapter 10 Grade Separations and Interchanges, Section 10.9.6.6.1 Tape-Type Exits, of AASHTO's *Green Book 7th edition* is referred to in this solution.

First determine the minimum deceleration lane length L_a using Table 10-6 along with the grade adjustments of Table 10-5, then add it to the length from the start of the taper to the location where the taper achieves a 12 ft wide wedge as shown below (drawing is Not to Scale):

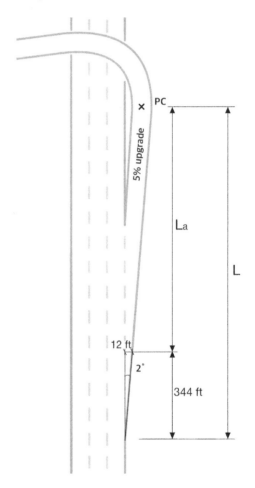

For more details on this refer to Figure 10-73 A – Tapered Design – Tangent.

Using Table 10-6 with the highway speed of $V = 70\ mph$ and a design speed of the controlling feature (which is PC in this case) of $V' = 30\ mph$:

$$L_a = 520\ ft$$

Using Table 10-5, with a 5% upgrade for a deceleration lane, the adjustment factor is '0.8'.

$$L_a' = 520\ ft \times 0.8 = 416\ ft$$

Finally, use trigonometry to determine the length of the tapered wedge that achieves a wedge width of $12\ ft$:

$$\frac{12\ ft}{\tan 2^o} = 343.6\ ft$$

Putting it altogether:

$$L = 343.6 + 416 \cong 760\ ft$$

Correct Answer is (A)

SOLUTION 3.44

Chapter 10 Grade Separations and Interchanges, Section 10.9.6.5.1 Taper-Type Entrances, of AASHTO's *Green Book 7th edition* is referred to in this solution.

Figure 10-72A is used along with the desirable rate of taper (50: 1 to 70: 1).

Using the rate of taper above along with a lane width at entrance of $16\ ft$, L is:

$$L = 16 \times 50\ \ to\ \ 16 \times 70$$
$$= 800\ ft\ \ to\ \ 1{,}120\ ft$$

Correct Answer is (D)

SOLUTION 3.45

Chapter 10 Grade Separations and Interchanges, Section 10.9.6.5.2 Parallel-Type Entrances, of AASHTO's *Green Book 7th edition* is referred to in this solution.

First determine the minimum acceleration length L_a using Table 10-4, then add to it the recommended length of taper which is $300\ ft$ as shown in Figure 10-72B of the Green Book.

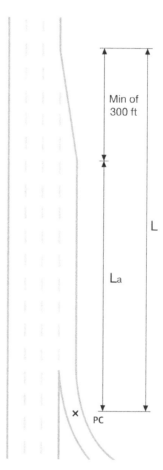

Using Table 10-4 with the highway speed of $V = 65\ mph$ and an average running speed at controlling feature on the ramp of $V_a' = 44\ mph$:

$$L_a = 370\ ft$$
$$L = 300 + 370 = 670\ ft$$

Correct Answer is (B)

SOLUTION 3.46

Chapter 3 Roadside Topography and Drainage Features, Section 3.2.4 Drainage Channels, of AASHTO's *Roadside Design Guide the 4th edition* is referred to in this solution.

Figure 3-7 from the above reference is used to determine the preferred cross section for this channel given that the width of its toe is $> 4\ ft$. Copied below for ease of reference (used with permission from AASHTO):

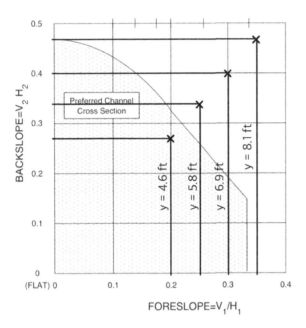

The preferred cross section should fall inside the shaded area of the graph. To determine this, Foreslopes and Backslopes for each proposed depth are calculated and plotted as shown above.

The below calculation is performed on option (A) which happens to be the only option that fits this criterion.

$$Foreslope = \frac{4.6}{23} = 0.2$$
$$Backslope = \frac{4.6}{17} = 0.27$$

Correct Answer is (A)

SOLUTION 3.47

Chapter 3 Roadside Topography and Drainage Features, Section 3.1 The Clear-Zone Concept, of AASHTO's *Roadside Design Guide the 4th edition* is referred to in this solution.

Use Table 3-1 along with the steepest of the two recoverable slopes, in which case this would be $1:8$. The clear-zone distance for this slope is $30\ ft$ to $34\ ft$.

Taking note (b) of Table 3-1 into account, and because recovery is unlikely on the $1:3$ slope, i.e., recovery of high-speed vehicles will occur beyond its toe. In other means, if a vehicle encroaches beyond the traveled way into the $1:8$ slope, it requires $30\ ft$ to $34\ ft$ to regain control. If this distance was not available, the balance of this distance should be provided at the toe of the non-recoverable $1:3$ slope (i.e., $3\ ft$ to $7\ ft$ because the $1:3$ begins $27\ ft$ beyond the $1:8$ slope as given in this question). However, the minimum in such cases is $10\ ft$ per Table 3-1(b). See below this information formulated in the following equation:

$$Runout = max((30\ ft\ to\ 34\ ft) - 27\ ft, 10\ ft)$$
$$= 10\ ft$$

See Figure 3-2 (*) of the reference for better clarity.

Correct Answer is (C)

(*) Figure 3-2 implies that if the non-recoverable slope was situated outside the clear-out distance, runout would not be necessary in this case.

SOLUTION 3.48

Chapter 3 Roadside Topography and Drainage Features, Section 3.1 The Clear-Zone Concept, of AASHTO's *Roadside Design Guide the 4th edition* is referred to in this solution.

Use Table 3-1 to determine the clear-zone distance for the inner and the outer Foreslopes, along with the given speed and traffic volume, as $20\ ft$ to $26\ ft$.

Use Table 3-2 to adjust the clear zone on the outside of the curvature as follows:

$$L_{C,outside} = 1.4 \times (20\ ft\ to\ 26\ ft)$$
$$= 28\ ft\ to\ 36.4\ ft$$

The clear zone for the inside of the curvature remains unadjusted:

$$L_{C,inside} = 20\ ft\ to\ 26\ ft$$

Putting it altogether:

$$L = L_{C,outside} + ramp + L_{C,inside}$$

$$L = (28\ to\ 36.4)\ ft + 12\ ft + (20\ to\ 26)\ ft$$

$$= 60\ ft\ to\ 74.4\ ft$$

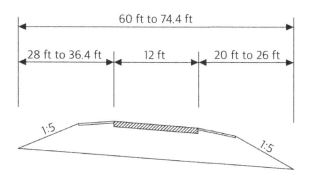

Option (A) with a total width of $65\ ft$ falls within the above identified range.

Correct Answer is (A)

SOLUTION 3.49

Chapter 5 Roadside Barriers, Section 5.6.1 Barrier Offset, of AASHTO's *Roadside Design Guide the 4th edition* is referred to in this solution.

Use Table 5-7 to determine the shy-line/barrier offset L_S for a $60\ mph$ design speed as $8\ ft$.

Correct Answer is (D)

SOLUTION 3.50

Chapter 5 Roadside Barriers, Section 5.6.4 Length-of-Need, of AASHTO's *Roadside Design Guide the 4th edition* is referred to in this solution.

Given that the barrier is not flared, Equation 5-2 can be used to determine the required length denoted as L – referred to as X in the following equation:

$$X = \frac{L_A - L_2}{L_A/L_R}$$

Refer to Figure 5-39 and Table 5-10(b) to determine the following Lengths:

$L_A = 9 + 7 + 7 = 23\ ft$

$L_2 = 9\ ft$

L_R is determined using Table 5-10(b) with a design speed of $70\ mph$ and traffic volume which is over $10,000\ veh/day$ as follows:

$L_R = 360\ ft$

$$L = X = \frac{23 - 9}{23/360} = 218.75\ ft$$

Correct Answer is (A)

SOLUTION 3.51

Chapter 5 Roadside Barriers, Section 5.6.2.1.1 Curb/Guardrail Combinations for Strong-Post W-Beam Guardrail, of AASHTO's *Roadside Design Guide the 4th edition* is referred to in this solution.

The above section states that roads with design speed lesser than 45 *mph*, guardrails should be either flushed with the curb, or placed no closer than 8 *ft* from the curb.

The 8 *ft* allows vehicles to return to their normal predeparture state/height which prevents them from jumping over the guardrail.

Correct Answer is (D)

SOLUTION 3.52

Although such treatment is not the preferred treatment, it can be used in locations where speed is 40 *mph* or less per Section 8.4.4.1 Sloped Concrete End Treatment of AASHTO's *Roadside Design Guide the 4th edition*.

Correct Answer is (B)

SOLUTION 3.53

Chapter 8 End Treatment, Section 8.4.3 Crash Cushions Based on Conservation of Momentum Principle – Table 8-9 with the sample design calculation, along with Equations 8-1 to 8-3 of AASHTO's *Roadside Design Guide the 4th edition* are all used in this solution.

$$V_1 = \frac{M_V V_o}{M_V + M_1}$$

$$= \frac{4{,}400 \; lb \times 90 \; mph}{4{,}400 \; lb + 700 \; lb}$$

$$= 77.65 \; mph$$

$$a = \frac{V_o^2 - V_1^2}{2D}$$

$$= \frac{[(1.47 \times 90) ft/sec]^2 - [(1.47 \times 77.65) ft/sec]^2}{2 \times 3 \; ft}$$

$$= 745.7 \; ft/sec^2$$

$$G = \frac{a}{g}$$

$$= \frac{745.7 \; ft/sec^2}{32.174 \; ft/sec^2}$$

$$= 23.2$$

Correct Answer is (C)

SOLUTION 3.54

Chapter 8 End Treatment, Tables 8-5, 8-6, and 8-7, of AASHTO's *Roadside Design Guide the 4th edition* are used in this solution.

Based on this, the following constitute self-restoring low maintenance crash cushions:

- ✓ **Compressor**
- ✓ **Quad Guard Elite**
- ✓ **Smart Cushion Innovations**

SOLUTION 3.55

Chapter 10 Roadside Safety in Urban or Restricted Environments, Section 10.1.3.2 Lane Merge Locations, of AASHTO's *Roadside Design Guide the 4th edition* is referred to in this solution.

The above section states that "the longitudinal placement of objects within approximately 10 *ft* of the taper point increases the frequency of roadside crashes...". Based on this:

$$L = 50 \times 12 + 10 = 610 \; ft$$

Correct Answer is (B)

SOLUTION 3.56

Chapter 4 Cross-Section Elements, Section 4.17.3 Curb Ramps, of AASHTO's *Green Book 7th edition* is referred to in this solution.

The above section states in page 4-70 the following in relevance to this question:

> "Cross slopes on adjacent sidewalks should be no greater than 2%".

This makes statement I incorrect.

> "A turning space at the top of each perpendicular curb ramp should be 4 ft by 4 ft."

This makes statement II true.

> "The maximum curb grade should be 8.33%"

This makes statement III incorrect.

> "Detectable warnings are needed where the curb has been removed to alert pedestrians with visual disabilities."

This makes statement IV true.

Correct Answer is (D)

SOLUTION 3.57

Chapter 8 Freeways, Section 8.2.10 Roadside Design, of AASHTO's *Green Book 7th edition* is referred to in this solution.

The second paragraph of this section states that the least distance in this case is 2 ft and this distance should be provided between the outer edge of a shoulder and the wall.

Correct Answer is (A)

SOLUTION 3.58

Several sections can be referred to in the AASHTO's *Green Book 7th edition*. To start with however, let's understand the intention behind traffic calming features.

Traffic calming is achieved with the use of physical changes, or designs, along with other measures to improve the overall safety of road users.

Considering this, only the following can be considered as traffic calming features (*):

- ✓ **Neighborhood roundabouts**
  For more details on this check page 9-27 of the Green Book.
- ✓ **Road humps**
  This is a physical change in the roadway which shall control and improve safety if applied correctly.
- ✓ **Diverging Diamond Interchange**
  For more details on this check page 10-53 of the Green Book.

(*) Law enforcement can be considered for traffic calming in some instances; however, geometric design can distract law enforcement personnel from achieving this target. Check Section 3.6.6 of the Green Book for more on this.

SOLUTION 3.59

AASHTO's *Green Book 7th edition*, page 4-66, the penultimate paragraph states that for individuals with disabilities the end of the walkway in this case should be ramped into the shoulder at a rate of 1: 20 – i.e., 5%.

Correct Answer is (A)

SOLUTION 3.60

AASHTO's *Green Book 7th edition*, page 5-8, Section 5.2.2.5 Bicycles and Pedestrian Facilities, states that such widened area should be provided at least every 200 ft.

Correct Answer is (C)

PART III
Design & Geometry

GEOTECHNICAL & PAVEMENT

Knowledge Areas Covered

SN	Knowledge Area
9	**Geotechnical and Pavement** A. Sampling, testing, evaluation, and soil stabilization techniques (e.g., soil classifications, subgrade resilient modulus, CBR, R-values, field tests, slope stability) B. Soil properties (e.g., strength, permeability, compressibility, phase relationships) C. Compaction, excavation, embankment, and mass balance D. Traffic characterization parameters, pavement design, and rehabilitation procedures (e.g., flexible and rigid pavement)

PART IV
Geotechnical & Pavement

PROBLEM 4.1 *Soil Moisture Content*
The maximum moisture content of a soil is close to its:

(A) Liquid limit

(B) Shrinkage limit

(C) Plastic limit

(D) Plasticity index

PROBLEM 4.2 *Settlement in Clay*
Time settlement in saturated clays when loaded, due to the addition of a building for example, is attributed to the following:

(A) Expulsion of clay particles

(B) The increase in effective stress of clay

(C) The deformation of clay particles

(D) All the above

PROBLEM 4.3 *Effective Stress Over Time*
The below embankment has a uniform weight of 1 ksf and was piled linearly over a period of 6 months on top of a layer of sand and clay as shown. The ground water level is 5 ft below the embankment. The density of sand and clay layers are both 120 pcf.

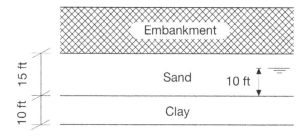

The profile that represents the change in effective stress over time at the bottom of each layer is:

(A) Profile A

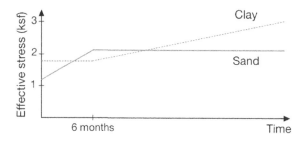

(B) Profile B

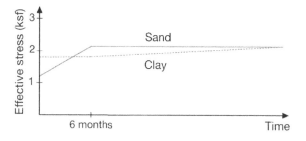

(C) Profile C

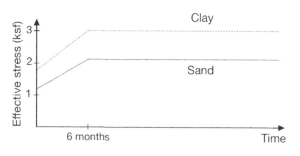

(D) Profile D

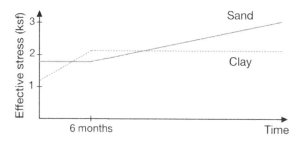

PROBLEM 4.4 *Soil Shear Strength*

The below is a particle distribution chart for two sands, Sand 1 (top) and Sand 2 (bottom).

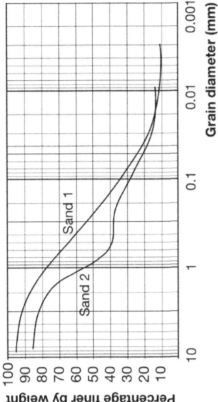

Based on this chart, the following statement is accurate during normal loading for the two sands:

(A) Sand 2 has a higher shear strength compared to Sand 1.

(B) Sand 1 has a higher shear strength compared to Sand 2.

(C) Both Sands have approximately the same shear strength.

(D) More information is required.

PROBLEM 4.5 *Subbase Stabilization*

A road contractor is requested to stabilize the subbase layer before the construction work starts as it includes moderates amount of clay gravel soils within its particles.

Considering that the *Plasticity Index* (PI) for the affected layer is '40', the best improvement method is as follows:

(A) Cement stabilization by mixing 3% Portland cement with the subbase material.

(B) Cement stabilization by mixing 9% Portland cement with the subbase material.

(C) Apply a small percentage of lime, typically 0.5% to 3%, to the affected material with a process called lime modification.

(D) Lime stabilization by mixing nearly 3% to 5% of lime to the affected layer.

PROBLEM 4.6 *Soil Properties*

The over consolidation ratio for a soil with a '0.33' normally consolidated at rest Rankine coefficient and a '0.85' over consolidated at rest coefficient is most nearly:

(A) 4.1

(B) 1.9

(C) 0.3

(D) 2.6

PROBLEM 4.7 *Soil Classification System (1)*

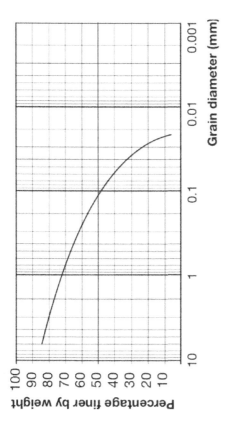

Using the Unified Soil Classification System, the above gradation sample is:

(A) GW

(B) GP

(C) SW

(D) SP

(✱) PROBLEM 4.8 *Soil Properties*

A soil sample that has a total volume of $1\ ft^3$ and a total mass of $100\ lb$ is removed from the ground. The water content of this sample is 20% and Specific Gravity '2.7'.

Based on the above information the following attributes are as follows:

The dry density of the sample is _____

The Degree of Saturation is _____

Porosity is _____

(✱) Normally you are asked to provide one value.

PROBLEM 4.9 *Soil Classification System (2)*

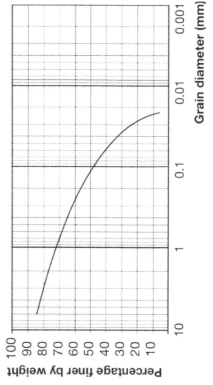

Based on the above soil sample gradation, and the following fines attributes:

- Liquid Limit (LL) = 50
- Plastic Limit (PL) = 35

Using the AASHTO classification system, the following represents the best group classification for this sample:

(A) A-7-5 (4)

(B) A-7-6 (4)

(C) A-7-4

(D) A-7-5

PROBLEM 4.10 *Soil Permeability Testing*

The following test is recommended for use to determine the coefficient of permeability for materials with lower permeability such as silts and clays:

(A) The constant head permeameter test

(B) The Piezometric test

(C) The falling head permeameter test

(D) The flexible wall permeameter test

PROBLEM 4.11 *Slope Stability/ Slope Safety Factor*

The below slope belongs to an excavation in a soil with cohesion $c = 58\ psf$ and density $\gamma = 95\ pcf$ along with a friction angle of $\emptyset = 20^o$.

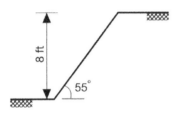

Using Taylor soil stability charts, the slope safety factor for this excavation is most nearly:

(A) 0.9

(B) 1.5

(C) 3.0

(D) 0.3

PROBLEM 4.12 *Optimum Moisture Content*

A proctor test was performed on *four* soil samples from the same batch using a proctor standard mold.

The weight of the moist samples after applying the test blows were: 4.32 lb, 3.92 lb, 3.92 lb and 4.36 lb, and their moisture content 11.6%, 17%, 8.8% and 14.8% respectively.

The optimum moisture content for this batch is most nearly:

(A) 14.0%

(B) 11.3%

(C) 12.5%

(D) 14.8%

PROBLEM 4.13 *Consolidation Settlement*

The below graph plots the results of an odometer test for a clay sample where x-axis represents the logarithm of pressure, and y-axis is the void ratio (e).

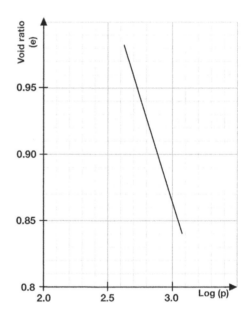

The expected settlement for a $4\ ft$ thick layer of this clay when pressure increases from an initial pressure of $500\ psf$ to a final pressure of $1,000\ psf$ is most nearly:

(A) 2.4 in

(B) 0.2 in

(C) 3.8 in

(D) 0.3 in

PROBLEM 4.14 Distresses in Flexible Pavements

The cause of a series of closely spaced ridges and valleys (ripples) in flexible pavements that are perpendicular to the traffic direction is usually caused by:

I. Insufficient base stiffness and strength.
II. Insufficient subgrade stiffness and strength.
III. Moisture and drainage problems.
IV. Freezing and thawing.

(A) I

(B) I + II

(C) I + II + III

(D) I + II + III + IV

PROBLEM 4.15 Rigid Pavement Cement

The preferred type of cement to be used in a Continuously Reinforced Concrete Pavement CRCP with the aim of reducing shrinkage is:

(A) Type III cement

(B) Type I cement

(C) Type II cement

(D) Type III mixed with fly ash

PROBLEM 4.16 Concrete Mix Design

A concrete mix with a specified w/c ratio of 0.45, a mix design of 1:1.5:3, assuming aggregate (*) is 5% in general moisture deficit to get into SSD, requires _____ liters of water to produce 1 ft^3 of yield.

(A) 5.2

(B) 6.2

(C) 10.5

(D) 19.3

(*) Assume density of 165 & 195 lb/ft^3 for aggregate and cement respectively.

PROBLEM 4.17 Soil Resistance Value

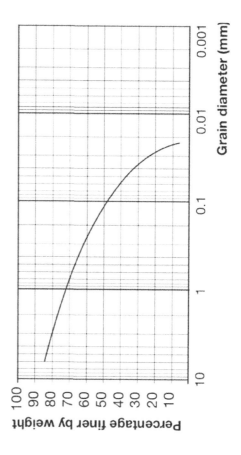

Based on the above soil sample gradation, and the following fines attributes:

- Liquid Limit (LL) = 50
- Plastic Limit (PL) = 35

The soil resistance value (R-Value) is most nearly:

(A) 35

(B) 26

(C) 20

(D) 53

PROBLEM 4.18 *Problem Soil*

Upon investigating a subgrade layer, the designer found the that the resilience modulus is low due to permanent high groundwater level.

Following are some strategies that could be implemented to improve this situation:

I. Remove water-soaked and weak soil and replace it with a new soil.
II. Place some separator layers that have high voids.
III. Increase the thickness of the base layers.
IV. Place a thick embankment for a period of time prior to construction to expel excess water.
V. Use geogrid to strengthen unbound layers.

(A) I + V

(B) II + III + IV

(C) III + IV + V

(D) II + III + V

PROBLEM 4.19 *Equivalent Single Axle Load ESAL for a Rigid Pavement*

A four-lane road (two lanes in each direction) with a forecasted traffic for the two directions of $AADT = 8,000 \; veh/day$ and the following traffic information:

- 8% two single axle truck: 8,000 lb/axle
- 4% three axle trucks, axle distribution per truck as follows:
  - Two single axles with 12,000 lb/axle
  - One tandem axle with 12,000 lb/axle
- Design life 30 *years*
- Growth rate 2% per year
- 80% of the traffic is in the design lane
- Rigid pavement 9 *in* thick
- Serviceability of 2.5

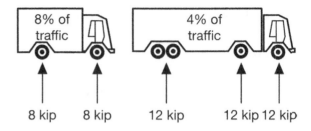

The Equivalent Single Axle Load ESAL for the design line and for the design life in one direction is most nearly:

(A) 42,789

(B) 959,042

(C) 1,243,343

(D) 1,830,617

PROBLEM 4.20 *Base Layer Thickness*

A Base layer's expected serviceability loss $\Delta PSI = 1.0$ along with an $ESAL = 5 \times 10^6$. Reliability $R = 90\%$ with a standard error $S_o = 0.3$. Resilience modulus $M_r = 10 \; ksi$, and elastic modulus $= 2 \times 10^5 \; psi$.

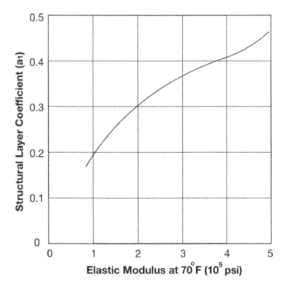

The thickness of this layer is most nearly:

(A) 10 *in*

(B) 15 *in*

(C) 20 *in*

(D) 25 *in*

SOLUTION 4.1

The *shrinkage limit* represents the water content that corresponds to transitioning between a brittle state and a semi-solid state.

The *plastic limit* represents the water content that corresponds to transitioning between a semi-solid state and a plastic state.

The *liquid limit* represents the water content that corresponds to transitioning from a plastic state to a liquid state.

The *plasticity index* represents the range within which soil remains in its plastic state bounded with its *plasticity limit* as the lower limit, and the *liquid limit* as the upper limit to this range.

This renders the *liquid limit* the maximum water content any soil can get to.

Correct Answer is (A)

SOLUTION 4.2

Clay is an undrained layer, which means that when loaded, water will not drain immediately. Rather, water, due to excess pressure, will drain/get expelled slowly and over a long period of time. The slow expulsion of water from the voids between clay particles causes the layer to lose its structure which leads into its ultimate and slow settlement over time.

When clay is loaded, and due its undrained property, pore water pressure increases. Upon water expulsion, pore water pressure decreases over time resulting in a gradual long-term increase in effective stress.

Correct Answer is (B)

SOLUTION 4.3

The effective stress σ' is the total stress σ removed from it the pore/water pressure u. In which case, the following effective stress changes at the bottom of each layer shall occur over the indicated period of 6 months and the age of the project/embankment.

Month zero prior to placing the embankment:
$$\sigma_{sand} = 120\ pcf \times 15\ ft$$
$$= 1{,}800\ psf\,(1.8\ ksf)$$

$$u_{sand} = 62.4\ pcf \times 10\ ft$$
$$= 624\ psf\,(0.62\ ksf)$$

$$\sigma'_{sand} = \sigma_{sand} - u_{sand}$$
$$= 1.8 - 0.62 \cong 1.2\ ksf$$

$$\sigma_{clay} = 120\ pcf \times (10 + 15)\ ft$$
$$= 3{,}000\ psf\,(3.0\ ksf)$$

$$u_{clay} = 62.4\ pcf \times 20\ ft$$
$$= 1{,}248\ psf\,(1.25\ ksf)$$

$$\sigma'_{clay} = \sigma_{clay} - u_{clay}$$
$$= 3.0 - 1.25 \cong 1.8\ ksf$$

Month 6 after placing the embankment:
When the embankment is placed, the sand layer will drain the excess (now pressurized) water immediately and hence no increase shall occur in the sand pore pressure. This will reflect in an increase in the effective pressure of the sand and given the loading from the embankment took place linearly over a period of 6 months, the increase in effective pressure for sand will be linear as well.

$$\sigma_{sand} = 0.12\ kcf \times 15\ ft + 1\ ksf$$
$$= 2.8\ ksf$$

$$u_{sand} = 62.4\ pcf \times 10\ ft$$

$$= 624 \, psf \, (0.62 \, ksf)$$

$$\sigma'_{sand} = \sigma_{sand} - u_{sand}$$
$$= 2.8 - 0.62 = 2.2 \, ksf$$

When it comes to the pore pressure of the clay layer, clay will not drain the excess (now pressurized) water right away (unlike sand). Drainage in this case shall occur over a long period of time instead. The pore pressure of the clay layer will increase due to this by the amount of the added load, and this keeps the effective stress unchanged.

$$\sigma_{clay} = 0.12 \, kcf \times 25 \, ft + 1 \, ksf$$
$$= 4.0 \, ksf$$

$$u_{clay} = 0.0624 \, kcf \times 20 \, ft + 1 \, ksf$$
$$= 2.25 \, ksf$$

$$\sigma'_{clay} = \sigma_{clay} - u_{clay} = 4.0 - 2.25$$
$$= 1.8 \, ksf$$

<u>Over a long period of time after placing the embankment:</u>
The sand effective stress will not change as the water has already drained from it long ago.

$$\sigma'_{sand} = \sigma_{sand} - u_{sand}$$
$$= 2.8 - 0.62$$
$$= 2.2 \, ksf$$

As for the clay layer, and over a long period of time, the excess (pressurized) water would have been drained then and this should bring the pore pressure down to:

$$u_{clay} = 0.0624 \, kcf \times 20 \, ft$$
$$= 1.25 \, ksf$$

$$\sigma'_{clay} = \sigma_{clay} - u_{clay}$$
$$= 4.0 - 1.25$$

$$= 2.8 \, ksf$$

This makes Profile A the most representative profile.

Correct Answer is (A)

SOLUTION 4.4

Generally, well graded Sands have higher friction angles compared to gap graded sands.

From the presented chart, Sand 2 seems to have gaps in its gradation around particle sizes $0.1 \, mm$ to $1.0 \, mm$, and this will have a detrimental effect on its friction angle.

Shear strength is proportional to cohesion and to the friction angle, see below:

$$\tau = c + \sigma_n tan\emptyset$$

τ Shear strength

c Total cohesion

σ_n Normal stress

$\emptyset$ Friction angle

It is therefore more likely that sand 1 will have a higher friction angle compared to Sand 2, which renders Sand 1 stronger in shear during normal loading.

Correct Answer is (B)

SOLUTION 4.5

Cement stabilization is considered when *plasticity index* is less than 10 and is used to strengthen granular soils by mixing in Portland cement, typically 3 – 5% of the soil dry weight. Check *NCEES Handbook* Section 2.5.4 for more information.

Lime modification is used to improve fine grained soils with the addition of 0.5% to 3% to the soil dry weight.

However, with a plasticity index > 10 for subbase or base materials with moderate and "predominant" amount of clay gravel soils, lime stabilization is considered the best option. Typically, 3% to 5% should be enough to dry up the mud contained in the subbase layer.

Correct Answer is (D)

SOLUTION 4.6
Reference is made to the *NCEES Handbook version 2.0*, Section 3.1.2.

At rest Rankine Coefficient for normally consolidated soils:

$K_{o,NC} = 1 - \sin\emptyset'$

$0.33 = 1 - \sin\emptyset' \rightarrow \sin\emptyset' = 0.67$

For over consolidated soils:
$K_{o,OC} = (1 - \sin\emptyset') \times OCR^{\sin\emptyset'}$

$= K_{o,NC} \times OCR^{\sin\emptyset'}$

$OCR = \left(\dfrac{K_{o,OC}}{K_{o,NC}}\right)^{\frac{1}{\sin\emptyset'}}$

$= \left(\dfrac{0.85}{0.33}\right)^{\frac{1}{0.67}}$

$= 4.1$

Correct Answer is (A)

SOLUTION 4.7
The following information is gathered from the gradation chart:

Nearly 45% is finer than 0.075 mm (No. 200 sieve), which means 55% is retained on this Sieve classifying the sample as either sand or gravel.

Nearly 82% is finer than 4.75 mm (No. 4 sieve), which means that 18% is retained on this sieve classifying the sample as sand.

$D_{10} = 0.025\ mm$

$D_{30} = 0.037\ mm$

$D_{60} = 0.27\ mm$

Coefficient of uniformity:

$C_u = D_{60}/D_{10} = 0.27/0.025 \approx 10$

Coefficient of curvature:

$C_c = (D_{30})^2/(D_{60}\ D_{10})$

$= (0.037)^2/(0.025 \times 0.27) \approx 0.2$

Based on the USCS classification system found in the *NCEES Handbook version 2.0* Table 3.7.2, for soils with > 50% retained on sieve No. 200, and > 50% passes No. 4 sieve, $C_u < 6$ and/or $C_c < 1$, soil would be classified as Poorly Graded Sand (SP).

Correct Answer is (D)

SOLUTION 4.8
Referring to the *NCEES Handbook version 2.0*, Section 3.8.3:

Dry Density:

$\gamma_d = \dfrac{\frac{W_t}{1+w}}{V}$

$= \dfrac{100lb/(1+0.2)}{1\ ft^3}$

$= 83.33\ lb/ft^3$

Degree of saturation:

$S = \dfrac{w}{\left(\dfrac{\gamma_w}{\gamma_d} - \dfrac{1}{G}\right)}$

$$= \frac{0.2}{\left(\frac{62.4}{83.33} - \frac{1}{2.7}\right)}$$

$$= 0.53$$

Porosity:

$$n = 1 - \frac{W_s}{GV\gamma_w} \left(= 1 - \frac{\gamma_d}{G\gamma_w}\right)$$

$$= 1 - \frac{83.33}{2.7 \times 62.4}$$

$$= 0.51$$

SOLUTION 4.9

Using the AASHTO classification system found in the *NCEES Handbook version 2.0* Section 3.7.3, nearly 45% is finer than 0.075 mm (i.e., passing No. 200 sieve). This indicates that the sample is not granular and falls within the Silt-Clay material categories of: A-4, 5, 6 or 7.

The characteristics of fines – material finer than 0.425 mm (i.e., passing No. 40 sieve) were given as follows:

$LiquidLimit(LL) = 50$

$Plasticlimit(PL) = 35$

$Plasticity\ Index\ (PI) = 50 - 35 = 15$

Based on this, the material is classified as either A-7-5 or A-7-6.

Using the comment section of the classification table and provided that $LL - 30 = 20 > PI$, classification of the material would be that of A-7-5.

The group index GI for this category is calculated as follows:

$GI = (F - 35)\ [0.2 + 0.005\ (LL - 40)]$
$\quad\quad + 0.01 \times (F - 15)(PI - 10)$

$= (45 - 35)\ [0.2 + 0.005\ (50-40)]$
$\quad\quad + 0.01 \times (45 - 15)(15 - 10)$

$= 4$

Final classification is A-7-5 (4)

Correct Answer is (A)

SOLUTION 4.10

The flexible wall permeameter test is used when the tested materials' permeability is lower than $1 \times 10^{-3} cm/sec$. The specimen in this case is encased in a membrane, and with the proper amount of pressure, flow through the specimen is recorded with time.

Correct Answer is (D)

SOLUTION 4.11

The *NCEES Handbook*, Chapter 3 Geotechnical, Section 3.6 Slope Stability is referred to.

The handbook provides two charts for Taylor (1948). The second is only applicable when friction angle $\emptyset = 0$ and a rock layer has been identified below the slope where the depth factor $D > 1$, which is not the case here.

The first chart – copied in the following page for ease of reference (used with permission from FHWA) – is used in this case.

There are two factors that we shall define prior to performing the calculation:

c_d is the developed, or mobilized, cohesion, which is the cohesion that develops at the slip surface upon failure.

ϕ_d is the developed, or mobilized friction angle, which is the friction angle that develops at the slip surface upon failure.

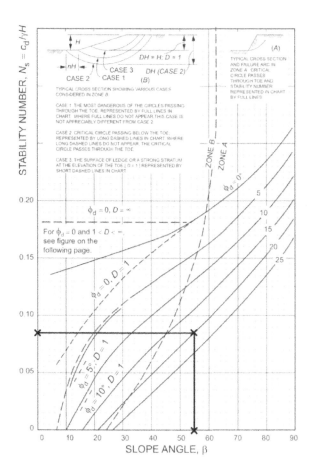

The safety factor requested in this question represents safety against forming a slip surface, in which case:

$$F.S. = F_c = \frac{c}{c_d}$$

Similarly, the safety factor for the friction angle shall be calculated and shall equal to:

$$F.S. = F_\phi = \frac{\tan \phi}{\tan \phi_d}$$

The above process is iterative in nature, and we may have to perform two or more iterations until the following equation is satisfied:

$$F.S. = F_\phi = F_c$$

Iteration 1: assume $\phi = \phi_d = 20°$

In reference to Taylor's (1948) first chart, using a slope angle $\beta = 55°$ – first iteration shown on the chart – the stability number $N_s = 0.085$.

$$N_s = \frac{c_d}{\gamma H}$$

$$\rightarrow c_d = \gamma H N_s$$
$$= 95 \, \frac{lb}{ft^3} \times 8 \, ft \times 0.085$$
$$= 64.6 \, psf$$

$$F_c = \frac{c}{c_d} = \frac{58 \, psf}{64.6 \, psf} = 0.9 = F_\phi$$

$$\phi_d = arctan\left(\frac{\tan \phi}{F_\phi}\right)$$
$$= arctan\left(\frac{\tan 20°}{0.9}\right)$$
$$= 22°$$

Iteration 2: assume $\phi_d = 22°$

In reference to Taylor (1948) chart, using interpolation $\rightarrow N_s = 0.077$.

$$c_d = \gamma H N_s$$
$$= 95 \, \frac{lb}{ft^3} \times 8 \, ft \times 0.077$$
$$= 58.5 \, psf$$

$$F_c = \frac{c}{c_d} = \frac{58 \, psf}{58.5 \, psf} = 0.99 = F_\phi$$

$$\phi_d = arctan\left(\frac{\tan \phi}{F_\phi}\right)$$
$$= arctan\left(\frac{\tan 20°}{0.99}\right)$$
$$= 20°$$

It is obvious at this stage that the safety factor falls somewhere between '0.9' to '1.0'.

Correct Answer is (A)

SOLUTION 4.12

The *NCEES Handbook,* Chapter 3 Geotechnical, Section 3.9 Laboratory and Field Compaction is referred to.

Optimum moisture content occurs at the soil's maximum dry density γ_d. In which case, dry density is calculated using the total (moist) density γ_t and moisture content w as follows:

$$\gamma_d = \frac{\gamma_t}{(1+w)}$$

Density is derived from the volume of the mold provided in the *Handbook* as $1/30\ ft^3$, based upon which, compaction curve is created as shown below:

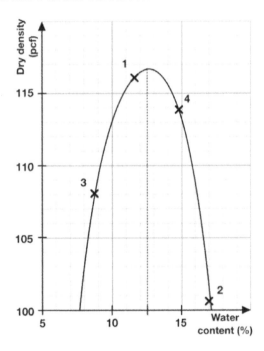

SN	wt_{moist} lb	γ_t lb/ft³	w %	γ_d lb/ft³
1	4.32	129.60	11.6%	116.13
2	3.92	117.60	17.0%	100.51
3	3.92	117.60	8.8%	108.09
4	4.36	130.80	14.8%	113.94

The optimum moisture content as derived from the curve is when dry density is max at 12.5%.

Correct Answer is (C)

SOLUTION 4.13

The *NCEES Handbook,* Chapter 3 Geotechnical, Section 3.2.1 Normally Consolidated Soils is referred to.

Settlement in a clay layer is calculated using the following equation:

$$S_C = \sum_{1}^{n} \frac{C_c}{1+e_o} H_o \, Log\left(\frac{p_f}{p_o}\right)$$

Where C_c is the compression index and is calculated using the slope of $'Log\,(p) - e'$ graph shown below (*). e_o is initial void ratio and can be picked up from the graph by substituting for $Log\,(p_o)$. H_o is the layer thickness, p_f is the final pressure and p_o is the initial/original pressure. n in the equation represents the number of layers, in which case the question did not specify more than *one* layer.

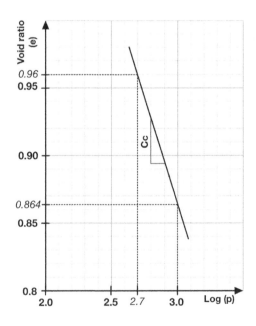

$$C_c = \frac{\Delta e}{\Delta \log(p)} \quad (*)$$
$$= \frac{0.96 - 0.864}{3.0 - 2.7}$$
$$= 0.32$$

$$e_o = e_{@\,[\log(500)\,=\,2.7]} = 0.96$$

$$S_C = \frac{C_c}{1+e_o} H_o \, Log\left(\frac{p_f}{p_o}\right)$$
$$= \frac{0.32}{1+0.96} \times 4\,ft \times Log\left(\frac{1{,}000}{500}\right)$$
$$= 0.2\,ft\,(2.4\,in)$$

Correct Answer is (A)

(*) For the removal of doubt, C_c is an absolute value and is calculated using the delta of void ratio Δe at the numerator. The graph shown in the *NCEES Handbook version 1.1* might be misleading and one may think that Δe should be at the denominator instead which is wrong, this is corrected in *version 1.2* and *2.0* now.

SOLUTION 4.14
The *NCEES Handbook,* Chapter 3 Geotechnical, Section 3.19 Pavements, is referred to.

The description given in the question is termed as Corrugation. Corrugation is caused by insufficient based stiffness and strength.

Correct Answer is (A)

SOLUTION 4.15
Type I & II are both used in transportation applications; however, type II is used when in contact with soils and groundwaters. Type II can also be applied to minimize ultimate shrinkage with proper controls for curing and water (see Page 173 of the MEPDG-3).

On the other hand, Type III cement is a high early strength development concrete. Its main use is in prestressed concrete and precast concrete applications. The addition of fly ash further increases its strength and reduces voids.

Correct Answer is (C)

SOLUTION 4.16
The weight of cement consumed to produce $1\,ft^3$ is calculated. Based upon which, the quantity of water can be determined based on the w/c ratio provided.

One bag of cement is 94 *lb* and is sufficient to produce the following yield:

Component	Ratio	Weight lb	Volume ft^3
Cement	1	94	0.48
Fine agg.	1.5	141	0.85
Coarse agg.	3	282	1.70
Water		42.3	0.68
TOTAL			3.71

Volume of water required for $1\,ft^3$

$= 0.68/3.71 = 0.183\,ft^3$ (5.18 *litres*)

Add an extra 5% to account for aggregates:

$= 5.4\,litres$

Correct Answer is (A)

SOLUTION 4.17
Reference is made in this question to the *Mechanistic-Empirical Pavement Design Guide MEPDG 3rd edition* Table 9-8.

Start with determining P_{200} which is the percent passing No. 200 Sieve (0.075 *mm*). Taken from the gradation graph as 45%.

From the characteristics of fines – material finer than $0.425\ mm$ (i.e., passing No. 40 sieve) were given as follows:

$$LiquidLimit(LL) = 50$$
$$Plasticlimit(PL) = 35$$
$$Plasticity\ Index\ (PI) = 50 - 35 = 15$$

Based on this, and in reference to Table 9-8 of the MEPDG (*):

$$CBR = \frac{75}{1 + 0.278\ (P_{200}\ PI)}$$
$$= \frac{75}{1 + 0.278\ (0.45 \times 15)}$$
$$= 26.1$$

$$M_r = 2555(CBR)^{0.64}$$
$$= 2555(26.1)^{0.64}$$
$$= 20,608.2\ psi$$

$$R = \frac{M_r - 1,155}{555}$$
$$= \frac{20,608.2 - 1,155}{555}$$
$$= 35.1$$

Correct Answer is (A)

(*) The same equations are also available in the AASHTO's Guide for Design of Pavement Structures GDPS Part I Chapter 1 page 14.

SOLUTION 4.18
Reference is made in this question to the *Mechanistic-Empirical Pavement Design Guide MEPDG 3rd edition*, Chapter 11 Pavement Design Strategies, Figure 11.1. The last item in Figure 11.1 applies to this question.

The recommended methods that apply to this problem soil are:

"II. Place some separator layers that have high voids."

"III. Increase the thickness of the base layers."

"V. Use geogrid to strengthen unbound layers."

Correct Answer is (D)

SOLUTION 4.19
Reference is made in this question to the *AASHTO's Guide for Design of Pavement Structures GDPS*.

The *NCEES Handbook* is referred to for growth factor calculation, as follows:

- NCEES Handbook growth factor of 2% over 30 *years* $(F/A, 2\%, 30) = 40.5681$

- GDPS Table D.13 for single axle, rigid pavement, and $P_t = 2.5$:
  - $LEF = 0.032$ for $8\ kip\ \&\ D = 9\ in$
  - $LEF = 0.176$ for $12\ kip\ \&\ D = 9\ in$

- GDPS Table D.14 for tandem axle, rigid pavement, and $P_t = 2.5$:
  - $LEF = 0.026$ for $12\ kip\ \&\ D = 9\ in$

- Per GDPS Appendix D-2, the directional distribution for traffic is 50%, i.e., $AADT = 4,000\ veh/day/direction$

ESAL for the 8% 2-single axle traffic:

$$ESAL = 2\ axles \times 8\% \times (4,000 \times 0.032) \times 80\% \times 365 \times 40.5681$$
$$= 242,603$$

ESAL for the 4% 2-single axle and 1-tandem axle traffic:

$ESAL_{single} = 2 \times 4\% \times (4{,}000 \times 0.176) \times 80\% \times 365 \times 40.5681$

$= 667{,}160$

$ESAL_{tandem} = 1 \times 4\% \times (4{,}000 \times 0.026) \times 80\% \times 365 \times 40.5681$

$= 49{,}279$

Adding them altogether:

$ESAL = 242{,}603 + 667{,}160 + 49{,}279$

$= 959{,}042$

Correct Answer is (B)

SOLUTION 4.20

Reference is made to the *AASHTO's Guide for Design of Pavement Structures GDPS* as follows:

- Part II, Chapter 2, Section 2.3.5 Layer Coefficient:

    This section defines the Structure Number SN, thickness D and the layer coefficient a which is taken from the figure in the question for base layer as $a_1 = 0.3$.

- Part II, Chapter 3, Table 3.1 – Design chart for Flexible Pavements – copied here for ease of reference (used with permission from AASHTO).

The below chart is used to determine the Structure Number SN.

To do this, will start from the left most of the nomograph using reliability $R = 90\%$ and extending a dotted arrow from this axis to the first turning line T_L passing through the standard deviation $S_o = 0.3$.

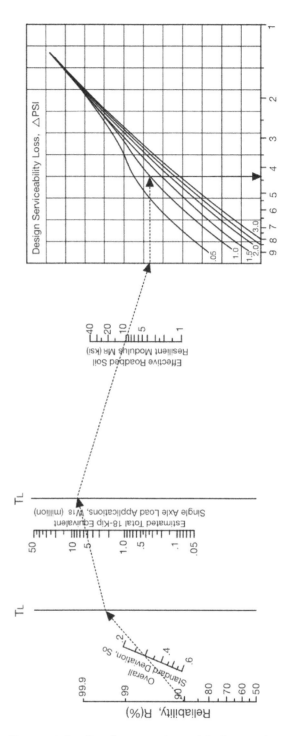

Connect the first intersection with the turning line T_L to the second turning line passing the dotted arrow through the ESAL axis at 5×10^6. From this point, connect the last arrow to the $\Delta PSI/SN$ chart passing through the M_r axis at 10 ksi.

Using the above chart, a Structure Number of $SN = 4.4$ is obtained for this layer.

Applying the layer coefficient equation per Section 2.3.5 gives us the requested thickness as follows:

$$SN = a_1 D$$
$$4.4 = 0.3 \times D$$
$$\rightarrow D = 14.7\ in$$

Correct Answer is (B)

DRAINAGE

Knowledge Areas Covered

SN	Knowledge Area
10	**Drainage** A. Hydrology, including runoff and water quality mitigation measures B. Hydraulics, including culvert and stormwater collection system design (e.g., inlet capacities, pipe flow, hydraulic energy dissipation, peak flow mitigation/detention, open-channel flow)

PART V
Drainage

PROBLEM 5.1 *Watershed Rainfall Depth*

The following watershed has a total area of 0.31 *Acre* and is plotted to scale on a 10 $ft \times 10\ ft$ grid.

Rain gauge stations 1, 2, 3 and 4 have been placed as shown and they measure the following rain depths:

Station 1 = 7.5 *in*

Station 2 = 5.5 *in*

Station 3 = 11.5 *in*

Station 4 = 7.5 *in*

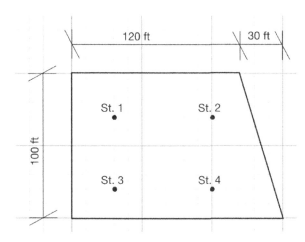

The average rainfall depth over the shown watershed using the Thiessen method is most nearly:

(A) 8.0 *in*

(B) 7.8 *in*

(C) 7.6 *in*

(D) 7.5 *in*

PROBLEM 5.2 *Precipitation Methods*

The most accurate method for averaging precipitation over an area is:

(A) The mathematical averaging method.

(B) The Isohyetal method.

(C) The Thiessen method.

(D) None of the above, each of those methods has a specific use and accuracy is irrelevant in this case.

PROBLEM 5.3 *Seawater Canal*

The below seawater return canal is made of concrete lining and is situated in an industrial area. Its purpose is to return seawater used to cool down equipment from factories in this industrial area back to the sea through an outfall.

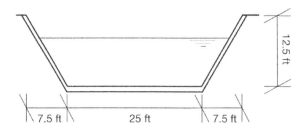

Assuming a continuous daily operation of this canal with a steady uniform flow, a design freeboard of 3.75 ft and a longitudinal slope of 2%, the seawater intake structure should be sized for a maximum intake (in MGD) of nearly:

(A) 9,000

(B) 7,750

(C) 5,300

(D) 9,200

PROBLEM 5.4 *Water Channel*

The below is a cross section of a V shaped open channel that delivers water at $70°F$ with a 1 ft free board. The mean velocity of the flow is 0.3 ft/sec.

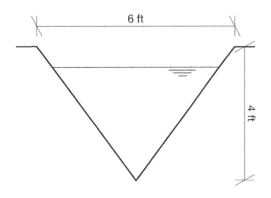

Reynolds number for this flow is most nearly:

(A) 0.32

(B) 32,000

(C) 0.26

(D) 26,000

(⁂) PROBLEM 5.5 *Water Discharge External Forces*

An inclined nozzle at 'B' with a diameter of 1.5 in, discharges 1 cfs of water into an open tank.

The supply pipe 'A' has a diameter of 3.5 in, and the nozzle is held in place by hinge 'C' as shown below.

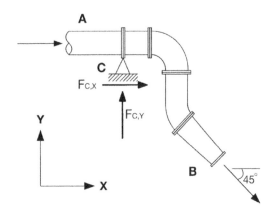

Ignoring the weight of water and the pipe arrangement, the horizontal and vertical reactions at support 'C' when pressure at 'A' is 100 psi is most nearly:

(A) $F_{C,x} = 754.5\ lb \leftarrow$
    $F_{C,y} = 236.75\ lb \downarrow$

(B) $F_{C,x} = 77.25\ lb \rightarrow$
    $F_{C,y} = 113.0\ lb \downarrow$

(C) $F_{C,x} = 837.0\ lb \leftarrow$
    $F_{C,y} = 176.7\ lb \downarrow$

(D) $F_{C,x} = 132.0\ lb \leftarrow$
    $F_{C,y} = 22.5\ lb \downarrow$

PROBLEM 5.6 *Head Losses*

The head loss in ft for a 20-year-old cast iron pipe, 200 ft long, 20 in dia, with a slope of 2%, is most nearly:

(A) 9

(B) 4

(C) 20

(D) 2

PROBLEM 5.7 *Unconfined Aquifer*

A 100 ft thick unconfined aquifer has a 12 in diameter well that pumps ground water from it at a rate of 65 gpm.

Assuming the radius of influence is 450 ft and permeability is $4 \times 10^{-4} ft/sec$, the drawdown at the well is most nearly:

(A) 96 ft

(B) 97.5 ft

(C) 4 ft

(D) 2.5 ft

(✱) PROBLEM 5.8 Elevation of Water Surface in Reservoir

A cast iron (*) pipeline with two 45° bends connects two large reservoirs as shown below. The diameter of the pipe is 12 in and is discharging 80° F water at a rate of 20 cfs.

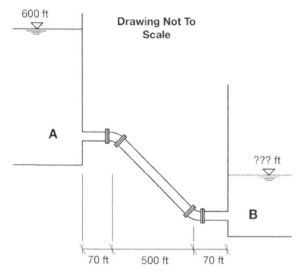

Given the elevations and distances in the above figure, elevation of the water surface at reservoir B is most nearly:

(A) 414 ft

(B) 190 ft

(C) 553 ft

(D) 50 ft

(*) Use lower range for roughness (ε) for cast iron pipes.

PROBLEM 5.9 Retention Pond Sizing

The size of a retention pond in ft^3 situated in a sloped zone that has 50 Acres of parks and cemeteries and the following data:

- Outflow rate of 35 cfs.
- Water flows in and out the pond in 15 minutes.
- Average rainfall intensity is 3.5 in/hr.

is most nearly:

(A) 7,875

(B) 131

(C) 39,375

(D) 31,500

PROBLEM 5.10 Travel Time for Shallow Flow

The below paved parking lot drains into the channel at its left side as shown. The parking has a slope of 0.5%.

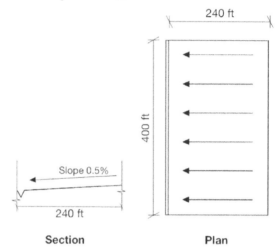

With a manning roughness of '0.011' and a 2-year 24-hour rainfall intensity of 2.0 in/hr. The travel time in minutes for a shallow flow over this plane to the channel is:

(A) 42.0 minutes

(B) 7.7 minutes

(C) 6.2 minutes

(D) 57.2 minutes

PROBLEM 5.11 Retention Basin Design

A slightly sloped 7 acres piece of land that used to be an old cemetery is to be fully redeveloped as follows:

- 45% downtown areas
- 35% playgrounds
- 15% asphalt roads
- 5% concrete walkways

Given the rainfall intensity in this area is $2\ in/hr$, the depth of a $250\ ft \times 150\ ft$ retention basin to be constructed to store the excess runoff from this redevelopment for one day is most nearly:

(A) $14\ ft$

(B) $12\ ft$

(C) $16.5\ ft$

(D) $18\ ft$

PROBLEM 5.12 *Shallow Flow*

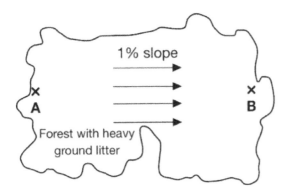

The above land is a forest characterized by heavy ground litter. This forest will be converted into short grass pasture. Land is sloped at a grade of $0.01\ ft/ft$.

The expected improvement in velocity of water flowing from point A to point B is most nearly:

(A) $0.74\ ft/sec$

(B) $0.25\ ft/sec$

(C) $0.50\ ft/sec$

(D) $1.00\ ft/sec$

PROBLEM 5.13 *Soil Loss Prevention*

The following bare land properties are provided for soil loss calculation and prevention:

- $300\ ft$ long sloped at 2%
- Type of soil is Sandy Loam with 0.5% organic matter
- Rainfall intensity index is $200\ ton.in/acre.yr$

Given a permissible soil loss limit of $11\ tons/hectare.yr$, the conservation factor should be:

(A) 0.3

(B) 0.74

(C) 0.014

(D) 0.033

PROBLEM 5.14 *Soil Erodibility*

Which of the following contributes to the amount of soil loss caused by erosion:

I. Porosity: porosity affects soil structure. The more porous the soil, the weaker its structure becomes, leading to increased erodibility.

II. Reduction in vegetation cover contributes to erosion.

III. An increase in kinetic energy from both wind and rainfall leads to higher erosion rates.

IV. Runoff distance plays a role. Longer runoff routes (i.e., longer lands) result in reduced erosion.

V. Runoff slope. The steeper the land slope, the greater the expected erosion.

(A) I + II + III + IV + V

(B) II + III + V

(C) I + II + III + V

(D) II + III + IV + V

PROBLEM 5.15 *Flow in a Gutter*

The below is a gutter with vertical curb located to the side of a road and is made of concrete (*).

The longitudinal slope of this gutter is 0.01 ft/ft.

The maximum flow this gutter can take during a storm event without the water overtopping the curb is most nearly:

(A) 6 cfs

(B) 36 cfs

(C) 1 cfs

(D) 235 cfs

(*) Use a manning coefficient for concrete of '0.013'.

PROBLEM 5.16 *Culvert Inlet Headwater Elevation*

A 100 ft long 33 in diameter concrete pipe culvert with a square/flushed cut end, having its inlet invert elevation at 317 ft, outlet invert elevation of 313 ft and a tailwater depth at outlet of 3 ft, has its <u>inlet</u> control's headwater elevation at _____ ft when subjected to a 40 cfs flow:

(A) 363 ft

(B) 321 ft

(C) 320 ft

(D) 325 ft

PROBLEM 5.17 *Culvert Outlet Headwater Elevation*

A 200 ft long 33 in diameter concrete pipe culvert with a square/flushed cut end, having its inlet invert elevation at 317 ft, outlet invert elevation of 313 ft and a tailwater depth at outlet of 3.5 ft, has its <u>outlet</u> control's headwater elevation at _____ ft when subjected to a 40 cfs flow:

(A) 318.6 ft

(B) 320.1 ft

(C) 322.6 ft

(D) 325.1 ft

PROBLEM 5.18 *Hydraulic Jump*

A hydraulic jump is occurring in an 8 ft wide rectangular channel with a discharge rate of 80 cfs and an initial water depth of 1 ft before the jump.

The depth of the water right after the hydraulic jump is most nearly:

(A) 3.0 ft

(B) 2.5 ft

(C) 2.0 ft

(D) 1.5 ft

PROBLEM 5.19 *Energy Loss*

The energy loss which occurs due to a hydraulic jump in a water channel from an initial depth of 1 ft to a final depth right after the jump of 4 ft is most nearly:

(A) 0.6 ft

(B) 1.7 ft

(C) 2.7 ft

(D) 3.0 ft

PROBLEM 5.20 *Pond Water Quality*

The following are effective strategy(s) that can safely treat large ponds for odor:

I. Aeration
II. Chemical treatment
III. Removing algae and other debris
IV. Bacterial treatment

(A) I

(B) II

(C) I + IV

(D) II + III

SOLUTION 5.1

The *Thiessen method* is a weighing method with a weight assigned to each of the gauging stations. Straight lines (the solid lines) are drawn to connect all the stations as shown in the figure below. Perpendicular bisectors (the dotted lines) of those connecting lines are then extended to form polygons around each station.

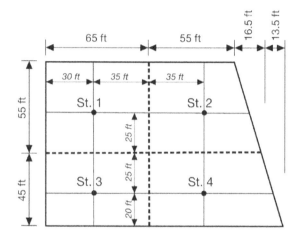

The area of each polygon is considered as the effective area per station.

A weighted average is then used to compute the required average rain depth.

St.	Effective area (A) ft^2	Gauge depth in	A × depth
1	3,575.00	7.5	26,812.50
2	3,478.75	5.5	19,133.13
3	2,925.00	11.5	33,637.50
4	3,521.25	7.5	26,409.38
	13,500.00		**105,992.50**

Average depth = 105,995.5/13,500 = 7.85 *in*

This method is referenced and explained in further detail in the *NCEES Handbook* Section 6.5.6.2 Thiessen Polygon Method.

Correct Answer is (B)

SOLUTION 5.2

The most accurate method for averaging precipitation over an area is the *Isohyetal method*. In this method, the amount of precipitation measured at each station is placed on the watershed map at gauges locations. Contours of equal precipitation, which are called Isohyets, are then drawn.

The area between Isohyets is determined accurately using planimetry. Based on those measures, the average precipitation for each area between those Isohyets is estimated by taking the average of two Isohyets sharing a boundary.

These values are then multiplied by each area's percentage and added together to obtain the weighted average.

The *NCEES Handbook version 2.0,* Section 6.5.6 Rainfall Gauging Stations, has a detailed elaboration for this method, the *Thiessen method,* and the *arithmetic averaging method* as well.

Correct Answer is (B)

SOLUTION 5.3

Manning Equation is used to solve this question.

$$Q = \frac{1.486}{n} A R_H^{2/3} S^{1/2}$$

Q is the discharge or flow rate (cfs).

A is the cross-sectional area of the flow (ft^2).

R_H is the hydraulic radius, this can be calculated by dividing the area of the flow (A) by the wetted Perimeter (P), or by using the tables provided in the *NCEES Handbook* Section 6.4.5.4.

S slope (ft/ft).

n is the manning roughness coefficient found in the *NCEES Handbook version 2.0* page 346 for concrete lined channel '0.015'.

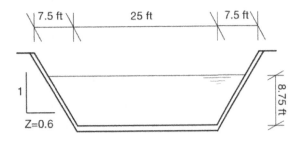

Using equations provided by *NCEES Handbook* and sourced by Chow (1959):

$$R_H = \frac{(b+zy)y}{b+2y\sqrt{1+z^2}}$$

$$= \frac{(25+0.6\times 8.75)\times 8.75}{25+2\times 8.75 \times \sqrt{1+0.6^2}}$$

$$= 5.83\ ft$$

$A = (b + zy)y$

$\quad = (25 + 0.6 \times 8.75) \times 8.75 = 264.7\ ft^2$

$Q = \frac{1.486}{0.015} \times 264.7 \times 5.83^{2/3} \times 0.02^{1/2}$

$\quad = 12{,}012.75\ ft^3/sec\ (= 7{,}764\ MGD)$

Correct Answer is (B)

SOLUTION 5.4
Reynold number R_e is calculated as follows:

$$R_e = \frac{vR_H}{\upsilon}$$

v is the mean velocity of the flow (given as $0.3\ ft/sec$).

υ is the kinematic viscosity for water which equals to $1.059 \times 10^{-5}\ ft^2/sec$ for water at $70°\ F$.

R_H is the hydraulic radius and equals to A/P which is calculated as follows for such a channel cross section (refer to the *NCEES Handbook* Section 6.4.5.4).

$$R_H = \frac{zy}{2\sqrt{1+z^2}}$$

$$= \frac{0.75 \times 3}{2\sqrt{1+0.75^2}}$$

$$= 0.9\ ft$$

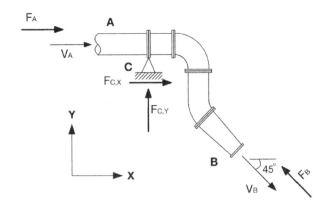

$$R_e = \frac{0.3 \times 0.9}{1.059 \times 10^{-5}}$$

$\quad = 25{,}496$

Correct Answer is (D)

SOLUTION 5.5
The impulse momentum principle is used to determine forces F_A and F_B which corresponds to the reactions at the support.

$$V_A = \frac{Q}{A_A} = \frac{1\,cfs}{\pi \times (1.75/12)^2\,ft^2} = 15.0\,ft/sec$$

$$V_B = \frac{Q}{A_B} = \frac{1\,cfs}{\pi \times (0.75/12)^2\,ft^2} = 81.5\,ft/sec$$

$$\sum F_x = \frac{\rho \times Q \times \Delta V_x}{g_c}$$

$$= \frac{\rho \times Q \times (V_{B,x} - V_{A,x})}{g_c}$$

$$= \frac{62.4\,pcf \times 1\,cfs \times (81.5\,cos(45) - 15)}{32.174\,lbm.ft/lbf.sec^2}$$

$$= 82.7\,lb$$

$$F_A - F_B \times cos(45) + F_{C,x} = 82.7\,lb$$

$$F_{C,x} = F_B \times cos(45) - F_A + 82.7\,lb$$
$$= P(A_B \times cos(45) - A_A) + 82.7\,lb$$
$$= 100(0.75^2\pi\,cos(45) - 1.75^2\pi) + 82.7$$
$$= -754.5\,lb \leftarrow$$

$$\sum F_y = \frac{\rho \times Q \times \Delta V_y}{g_c}$$

$$= \frac{\rho \times Q \times (V_{B,y} - V_{A,y})}{g_c}$$

$$= \frac{62.4\,pcf \times 1\,cfs \times (-81.5\,sin(45) - 0)}{32.174\,lbm.ft/lbf.sec^2}$$

$$= -111.8\,lb$$

$$F_B \times sin(45) + F_{C,y} = -111.8\,lb$$

$$F_{C,y} = -111.8\,lb - F_B \times sin(45)$$
$$= -111.8\,lb - P \times A_B \times sin(45)$$
$$= -111.8\,lb - 100\,psi \times 0.75^2\pi\,sin(45)$$
$$= -236.75\,lb \downarrow$$

Correct Answer is (A)

SOLUTION 5.6
Using the Hazen-Williams equation and coefficients provided in Section 6.3 of the *NCEES Handbook version 2.0*, see below:

$$h_f = \frac{4.73\,L}{C^{1.852}\,D^{4.87}}\,Q^{1.852}$$

C is the Hazen-Williams Coefficient which equals to 100 for 20-year-old pipes. L and D are length and diameter in ft.

$$Q = 1.318\,C\,A\,R_H^{0.63}\,S^{0.54}$$

R_H is the hydraulic radius, and it equals to the area of flow (A) divided by the wetted perimeter (P). For pipelines running in full capacity, $R_H = r/2$

$$Q = 1.318 \times 100 \times \pi \times \left(\frac{10}{12}ft\right)^2 \times \left(\frac{5}{12}ft\right)^{0.63} \times 0.02^{0.54}$$

$$= 20\,ft^3/sec$$

$$h_f = \frac{4.73 \times 200}{100^{1.852} \times \left(\frac{20}{12}\right)^{4.87}} \times 20^{1.852}$$

$$= 4.0\,ft$$

A quicker way for determining the answer is by using the slope S as follows:

$$S = h_f/L$$

$\rightarrow h_f = S \times L = 0.02 \times 200 = 4\,ft$

Correct Answer is (B)

SOLUTION 5.7
The *NCEES Handbook*, Chapter 6 Water Resources and Environmental, Dupuit's Formula can be used to solve this question.

$$Q = \frac{\pi K (h_2^2 - h_1^2)}{\ln\left(\frac{r_2}{r_1}\right)}$$

Q is the flow rate in ft^3/sec, h_1 and h_2 are heights of the aquifer measured from its bottom at the perimeter of the well (i.e., $r_1 = \frac{12}{2}\ in = 0.5\ ft$) and at the influence radius of $r_2 = 450\ ft$ respectively.

Radius of influence defines the outer radius of the cone of depression, hence $h_2 = 100\ ft$.

Check the below figure for more clarity:

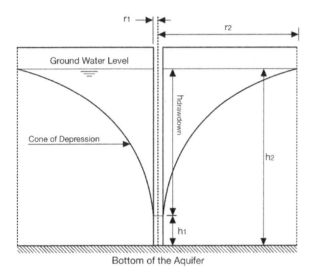

$$h_1 = \sqrt{\frac{\pi K h_2^2 - Q \times \ln\left(\frac{r_2}{r_1}\right)}{\pi K}}$$

$$= \sqrt{\frac{\pi \times 4 \times 10^{-4} \frac{ft}{sec} \times (100\ ft)^2 - 65 \frac{gal}{min}\left(\frac{0.134\ ft^3}{gal}\right)\left(\frac{1\ min}{60\ sec}\right) \times \ln\left(\frac{450\ ft}{0.5\ ft}\right)}{\pi \times 4 \times 10^{-4} \frac{ft}{Sec}}}$$

$$= 96\ ft$$

$h_{drawdown} = 100\ ft - 96\ ft = 4\ ft$

Correct Answer is (C)

SOLUTION 5.8

The *NCEES Handbook,* Section 6.3 Closed Conduit Flow and Pumps can be used to solve this problem.

Given the inputs in this question, the following equations/sections will be referred to:

- The Energy Equation – Section 6.2.1.2
- Reynolds Number for circular pipes – Section 6.2.2.2
- Darcy-Weisbach Equation for head losses due to flow – Section 6.2.3.1
- Minor Losses in Pipe Fittings, Contractions and Expansions – Section 6.3.3

The Energy Equation:

$$\frac{P_A}{\gamma} + z_A + \frac{v_A^2}{2g} = \frac{P_B}{\gamma} + z_B + \frac{v_B^2}{2g} + h_f$$

Pressure at surface A and surface B is zero (atmospheric only). Also, velocity of the surface dropping in level can be neglected in large reservoirs.

With this, the energy equation can be reduced to:

$$z_A = z_B + h_f$$

Calculating Head Losses h_f:

Head loss due to flow:

Using Darcy-Weisbach Equation for head losses:

$$h_{f,flow} = f \frac{L}{D} \frac{v^2}{2g}$$

f is a function of both Reynolds Number.

R_e and the relative roughness $\left(\frac{\varepsilon}{D}\right)$ taken from the Moody, Darcy, or Stanton Friction Factor Diagram page 312 of the *NCEES Handbook* version 2.0.

v in the denominator of Reynolds Number equation (shown below) is the kinematic viscosity taken from the physical properties of water table Section 6.2.1.6 as $0.93 \times 10^{-5} \; ft^2/sec$.

v in the numerator is velocity in the pipe

$$= \frac{Q}{A} = \frac{20 \; ft^3/sec}{\pi (0.5 \; ft)^2} = 25.5 \; ft/sec$$

$$R_e = \frac{vD}{v}$$

$$= \frac{25.5 \; ft/sec \times 12 \; in \times \frac{1 \; ft}{12 \; in}}{0.93 \times 10^{-5} \; ft^2/sec}$$

$$= 2.7 \times 10^6$$

$$\frac{\varepsilon}{D} = \frac{0.0006}{1 \; ft} = 0.0006$$

Substitute R_e and $\frac{\varepsilon}{D}$ in Moody, Darcy, or Stanton Friction Factor Diagram page 312 of the *NCEES Handbook version 2.0* generates a friction value of $f = 0.019$

$$h_{f,flow} = f \frac{L}{D} \frac{v^2}{2g}$$

$$= 0.019 \times \frac{\left(70 + \frac{500}{\sin(45)} + 70\right) ft}{1 \; ft} \frac{\left(25.5 \frac{ft}{sec}\right)^2}{2 \times 32.174 \frac{ft}{sec^2}}$$

$$= 162.6 \; ft$$

Head loss due to pipe fittings:

$$h_{f,fittings} = C \frac{v^2}{2g}$$

Where C equals to '1.0', '0.5' and '0.4' for sharp reservoir exit, sharp reservoir entrance and 45° elbow respectively taken from Sections 6.3.3.1 and 6.3.3.2.

$$h_{f,fittings} = C_{total} \frac{v^2}{2g}$$

$$= (1 + 2 \times 0.4 + 0.5) \times \frac{\left(25.5 \frac{ft}{sec}\right)^2}{2 \times 32.174 \frac{ft}{sec^2}}$$

$$= 23.2 \; ft$$

Energy Equation and Total Head Losses:

$$z_B = z_A - h_f$$
$$= z_A - (h_{f,flow} + h_{f,fittings})$$
$$= 600 - (162.6 + 23.2)$$
$$= 414.2 \; ft$$

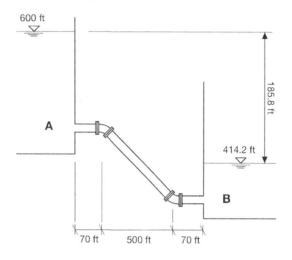

Correct Answer is (A)

SOLUTION 5.9

Referring to Section 6.5.9.1 of the *NCEES Handbook*, the size of a retention pond using the rational method is calculated as follows – considering that the runoff coefficient C for sloped parks and cemeteries (check comment below the NCEES runoff table) is '0.25' per Section 6.5.2.1:

$$V_s = V_{in} - V_{out}$$
$$= (i \sum AC - Q_o) \times t$$
$$= \left(3.5 \frac{in}{hr} \times 50 \; Acres \times 0.25 - 35 \; cfs\right) \times 15 \; min \times 60 \frac{sec}{min}$$
$$= 7,875 \; ft^3 \; (*)$$

Correct Answer is (A)

(*) Although the *NCEES Handbook version 2.0* identify 't' as time in minutes, given the units consistency in this question, 't' is measured in seconds.

SOLUTION 5.10

This is a sheet flow, which is a flow over plane surfaces at very shallow depths (about $0.1\ ft$). Per *NCEES Handbook*, Section 6.5.4.2, sheet flow travel time in minutes over a flow length $L\ (ft)$, and a slope S measured in (ft/ft) is calculated as follows:

$$T_{ti} = \frac{K_u}{I^{0.4}} \left(\frac{nL}{\sqrt{S}}\right)^{0.6}$$

$$= \frac{0.933}{2.0^{0.4}} \left(\frac{0.011 \times 240\ ft}{\sqrt{0.005}}\right)^{0.6} = 6.2\ minutes$$

Correct Answer is (C)

SOLUTION 5.11

The *NCEES Handbook version 2.0*, Section 6.5.2 Runoff Analysis can be used. Given the input provided in the question, the best method to be implemented is the Rational Method.

$$Q = CIA$$

Q is the discharge in ft^3/sec, this value is calculated pre and post development, the difference of both shall be used to design the retention basin.

C is the runoff coefficient which can be obtained from the *NCEES Handbook*. Given the area is only slightly sloped, the lower range of C in the runoff table can be used.

I is rainfall intensity given in the question as $2\ in/hr$, and A is area which is $7\ acres$.

Pre-development:

For slightly sloped parks and cemeteries $C = 0.10$

$$Q_{pre} = CIA$$
$$= 0.1 \times 2\ in/hr \times 7\ acres$$
$$= 1.4\ ft^3/sec$$

$$V_{pre} = Q_{pre} \times 1\ day$$
$$= 1.4\ ft^3/sec \times 86{,}400\ sec\ per\ day$$
$$= 120{,}960\ ft^3/day$$

Post-development:

Calculate weighted runoff as follows:
- Downtown areas → $C = 0.70$
- Playgrounds → $C = 0.20$
- Asphalt roads → $C = 0.70$
- Concrete walkways → $C = 0.80$

$$C_w = 45\% \times 0.7 + 35\% \times 0.2$$
$$+ 15\% \times 0.7 + 5\% \times 0.8$$
$$= 0.53$$

$$Q_{post} = CIA$$
$$= 0.53 \times 2\ in/hr \times 7\ acres$$
$$= 7.42\ ft^3/sec$$

$$V_{post} = Q_{post} \times 1\ day$$
$$= 7.42\ ft^3/sec \times 86{,}400\ sec\ per\ day$$
$$= 641{,}088\ ft^3/day$$

$$\Delta V = V_{post} - V_{pre}$$
$$= 641{,}088 - 120{,}960$$
$$= 520{,}128\ ft^3$$

The depth of the rectangular pond is therefore calculated as follows:

$$d = \frac{520{,}128\ ft^3}{250\ ft \times 150\ ft} = 13.87\ ft$$

Correct Answer is (A)

SOLUTION 5.12

The Water Velocity Versus Slope for Shallow Concentrated Flow diagram page 397 of the *NCEES Handbook version 2.0*, Section 6.5.5 Hydrograph Development and Applications can be used to solve this problem.

The graph is pasted below for ease of reference.

A horizontal line is constructed from the y-axis at 0.01 which intersects with the desired two velocity graphs as shown:

$$v_{forest} = 0.25 \, ft/sec$$

$$v_{short\ grass} = 0.74 \, ft/sec$$

$$\Delta v = 0.74 - 0.25 = 0.49 \, ft/sec$$

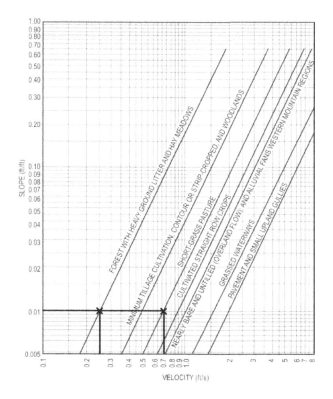

Correct Answer is (C)

SOLUTION 5.13

The *NCEES Handbook version 2.0*, Chapter 6 Water Resources and Environmental, Section 6.5.9.2 Erosion/ Revised Universal Soil Loss Equation of page 411 is referred to.

$$A = R.K.LS.C.P$$

P is the conservation factor, given all inputs of this equation are provided in the body of the question, P is calculated as follows:

$$P = \frac{A}{R.K.LS.C}$$

A is the amount of soil loss due to erosion measured in *tons per acre per year*:

$$A = 11 \, \frac{tons}{hectare.yr} \times \frac{1\ hectare}{2.47\ acre}$$

$$= 4.45 \, \frac{tons}{acre.yr}$$

K is the soil erodibility factor taken from the same section of *NCEES Handbook version 2.0* page 412 as '0.27' for Sandy Loam with 0.5% organic matter.

LS is the topographic factor taken from the same section of *NCEES Handbook version 2.0* page 412 as '0.28' for a 300 ft land sloped at 2%.

C is the crop and cover management factor taken as '1.0' for bare land.

$$P = \frac{A}{R.K.LS.C}$$

$$= \frac{4.45}{200 \times 0.27 \times 0.28 \times 1.0}$$

$$= 0.29 \, (*)$$

Correct Answer is (A)

(*) A conservation value of $P = 0.29$ requires strip cropping and contour farming. In a nutshell this requires growing crops in a systematic arrangement of strips along the contours and across a sloping field.

SOLUTION 5.14

The *NCEES Handbook version 2.0,* Chapter 6, Section 6.5.9.2 Erosion/ Revised Universal Soil Loss Equation is referred to in order to provide context into the solution.

The Revised Universal Soil Loss Equation:

$$A = R.K.LS.C.P$$

A is the amount of soil loss due to erosion (*tons per acre per year*), R is the rainfall erosion index or the climatic erosivity, K is soil erodibility factor, LS is the topographic factor and is taken from the table provided in page 412 of the handbook's version 2.0. C represents vegetation and is called the crop and cover management factor, and P is the erosion control practices factor.

Factors Influencing Soil Loss:

Rainfall erosion index R considers the intensity, duration, and continuity of rainfall. Rainfall erosivity and its relationship to kinetic energy play a crucial role in erosion.
This makes Statement III true.

Soil erodibility K is its susceptibility to erosion. Factors affecting K include soil aggregation and structure. The more porous the soil, the reduced runoff it shall experience and the lesser effect it would have on its continuous erodibility.
This marks Statement I as incorrect.

In a similar fashion, vegetation cover acts as a barrier against erosion by obstructing water velocity, hence lesser runoff is experienced with more cover.
This makes Statement II true.

Topographic factor LS, specifically the slope length and its steepness, are both proportional to erosion. Which means, more length and more slope causes more erosion.
This marks Statement IV as incorrect, however, it makes Statement V true.

Correct Answer is (B)

SOLUTION 5.15

The curb and gutter equation from Section 6.4.10.3 of the *NCEES Handbook* is used to solve this problem.

$$Q = \left(\frac{0.56}{n}\right) \frac{S_L^{0.5}}{S_x} d^{8/3}$$

Q is the discharge or flow rate (cfs).

S_x is the cross slope in which case is $0.12 \, ft/ft$ and S_L is the longitudinal slope being $0.01 \, ft/ft$.

n is the manning roughness coefficient found in the *NCEES Handbook version 2.0* for concrete as '0.013'.

$$Q = \left(\frac{0.56}{0.013}\right) \frac{0.01^{0.5}}{0.12} 0.5^{8/3}$$
$$= 5.7 \, cfs$$

Correct Answer is (A)

SOLUTION 5.16

Reference is made to the *Hydraulic Design of Highway Culverts FHWA HIF-12-026,* Appendix C, Chart 1B – Headwater Depth For Concrete Pipe Culverts with Inlet Control. The chart is pasted on the following page for ease of reference (used with permission from FHWA). The same chart is available in the *NCEES Handbook version 2.0* page 370.

A straight line is drawn from the first scale that represents the diameter of the culvert.

The line starts at a diameter of 33 in as given in the question, and it passes through a flow of 40 cfs in the following scale which determines the value of HW/D in the first of the three scales to the rightmost of the graph, which represents a square edge with the headwall:

$$HW/D = 1.4$$

With a diameter $D = 33\ in$

$$HW = 1.4 \times 33 = 46.2\ in\ (= 3.85\ ft)$$

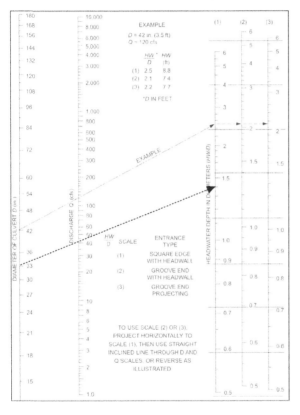

$Elevation\ of\ HW = 317 + 3.85 = 320.85\ ft$

Correct Answer is (B)

SOLUTION 5.17
Reference is made to the *Hydraulic Design of Highway Culverts FHWA HIF-12-026*, Appendix C, Chart 5B – Head for Concrete Pipe Culverts Flowing Full (used with permission from FHWA). The same chart is also available in the *NCEES Handbook version 2.0* page 373.

The entrance loss coefficient K_e shall be determined first for use in the said chart. This coefficient can be taken from Table C.2 of Appendix C of the FHWA manual referenced above, or the *NCEES Handbook version 2.0* page 369 – taken as '0.5' for projected concrete pipes with a cut/square-edge end.

Using the graph copied below, start with the dotted line shown. The dotted line connects the diameter scale at 33 in and the length scale of $K_e = 0.5$ at 200 ft. An **x** is marked on the turning line for later use.

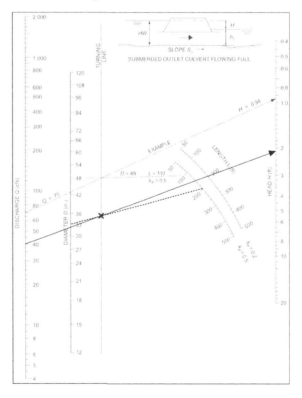

A solid line/arrow is then extended starting from a discharge of 40 cfs, passing through the **x** marked previously, and points towards the last scale (the head) which reads 2.1 ft in this case. The above information is valid as long as $0.5 < HW/D < 3.0$. This information can be verified later with the help of the following sketch:

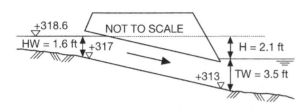

The requested water elevation at inlet can be determined graphically, considering the level at the inlet (+317) and the level at the outlet (+313), or by using the following equation taken from the mentioned FHWA publication Equation 3.6 b page 3.11:

$$HW = TW + H - LS$$

TW is the tailwater depth ($3.5\,ft$), H is the head calculated previously ($2.1\,ft$) and LS is the drop in elevation ($4\,ft$).

$$HW = 3.5 + 2.1 - 4 = 1.6\,ft$$

$$\frac{HW}{D} = \frac{1.6\,ft}{33\,in \times \frac{1\,ft}{12\,in}} = 0.58 \rightarrow ok$$

$$Elevation\ of\ HW = 317 + 1.6 = 318.6\,ft$$

Correct Answer is (A)

SOLUTION 5.18

The *NCEES Handbook version 2.0,* Chapter 6 Water Resources and Environmental, Section 6.4.3.1 Froude Number, and Section 6.4.8.1 Depths and Flows, are both referred to in this solution.

Calculate Froude number for use in subsequent equations per Section 6.4.3.1 as follows:

$$Fr_1 = \frac{v}{\sqrt{gy_1}}$$

$$v = \frac{Q}{A} = \frac{80\,ft^3/sec}{8\,ft \times 1\,ft} = 10\,ft/sec$$

$$g = 32.2\,ft/sec^2$$

$$Fr_1 = \frac{10\,ft/sec}{\sqrt{32.2\,ft/sec^2 \times 1\,ft}} = 1.76\,(*)$$

$$\frac{y_2}{y_1} = \frac{1}{2}\left(\sqrt{1 + 8Fr_1^2} - 1\right)$$

$$\frac{y_2}{y_1} = \frac{1}{2}\left(\sqrt{1 + 8 \times (1.76)^2} - 1\right) = 2.0$$

$$y_2 = 2.0 \times 1 = 2.0\,ft$$

Another shorter method can be used by applying the following equation found on page 354 of the *NCEES Handbook version 2.0*:

$$y_2 = -\frac{1}{2}y_1 + \sqrt{\frac{2v_1^2 y_1}{g} + \frac{y_1^2}{4}}$$

$$= -\frac{1}{2}(1) + \sqrt{\frac{2(10)^2(1)}{32.2} + \frac{(1)^2}{4}}$$

$$= 2.0\,ft$$

Correct Answer is (C)

(*) In reference to Section 6.4.8.2, a Froude number which falls between '1.0' and '1.7' generates undular/low energy jumps.

SOLUTION 5.19

The *NCEES Handbook version 2.0,* Chapter 6 Water Resources and Environmental, Section 6.4.8.6 Energy Loss in Horizontal Hydraulic Jump is referred to in this solution.

With the lack of information about velocity, the following equation can be used:

$$\Delta E = \frac{(y_2 - y_1)^3}{4y_1 y_2} = \frac{(4-1)^3}{4(4)(1)} = 1.7\,ft$$

Correct Answer is (B)

SOLUTION 5.20

Odor in water bodies can arise from various sources, with the most prevalent being the overgrowth of algae. When algae die and decompose, they deplete the oxygen supply, hindering the breakdown of organic waste, consequently leading to unpleasant odors.

Water stagnation is another contributing factor to foul odors. When water remains still, algae and bacteria grow, and these bacteria break down waste materials, generating carbon dioxide and hydrogen sulfide, resulting in an odor similar to rotten eggs.

Additionally, the accumulation of scum at the bottom of ponds, which is mostly contributed by runoffs, can also cause water to stink.

Effective strategies to address these issues are:

○ Aeration: reintroducing oxygen into the pond to prevent stagnation. This can be accomplished through aeration systems like fountains and submerged aeration systems.

○ Specialized Bacterial Blends: The introduction of specific bacterial blends capable of mitigating algae growth through denitrification is beneficial. Stormwater ponds typically contain elevated levels of nitrogen and phosphorous, which serve as nutrients for algae. Specially designed bacterial treatments can digest these nutrients before algae can, thus impeding their growth. Moreover, these bacteria can reach the pond's bottom to break down the sludge layer formed by runoff, reducing odor.

The above makes statements I and IV true.

Apart from the above, it is important to note that chemical treatments can also address sludge and algae issues, as well as reduce nitrogen and phosphorus levels. However, the use of chemicals should be considered a last resort due to their potential adverse effects on the environment, aquatic life, and other ecologically sensitive elements.

Moreover, while physically removing algae and debris from ponds is another viable strategy, it tends to be a short-term solution, especially for large ponds, and may entail substantial costs. Furthermore, this approach does not address the underlying causes of the problem.

The above marks statements II and III as incorrect.

Correct Answer is (C)

PART V
Drainage

REFERENCES

References

Code	Title	Institution
HCM	Highway Capacity Manual (Volumes 1-4), 6th edition, 2016 www.mytrb.org	Transportation Research Board, National Research Council, Washington, D.C.
MUTCD	Manual on Uniform Traffic Control Devices for Streets and Highways, 2009, including Revisions 1 and 2 dated May 2012 www.mutcd.fhwa.dot.gov	U.S. Department of Transportation, Federal Highway Administration, Washington, D.C
AASHTO HSM-1	Highway Safety Manual, 1st edition, 2010, with 2014 Supplement (including September 2010, February 2012, and March 2016 errata) www.tansportation.org	American Association of State Highway & Transportation Officials, Washington, D.C.
FHWA HIF-12-026	Hydraulic Design of Highway Culverts, Hydraulics Design Series Number 5, 3rd edition, April 2012 www.fhwa.dot.gov	U.S. Department of Transportation, Federal Highway Administration, Washington, D.C
AASHTO GDHS-7	A Policy on Geometric Design of Highways and Streets 7th edition, 2018 (including October 2019 errata) www.tansportation.org	American Association of State Highway & Transportation Officials, Washington, D.C
AASHTO GPF-2	Guide for the Planning, Design, and Operation of Pedestrian Facilities, 2nd edition, 2021 www.tansportation.org	American Association of State Highway & Transportation Officials, Washington, D.C
AASHTO RSDG-4	Roadside Design Guide, 4th edition, 2011 (including February 2012 and July 2015 errata) www.tansportation.org	American Association of State Highway & Transportation Officials, Washington, D.C
AASHTO MEPDG-3	Mechanistic-Empirical Pavement Design Guide: A Manual of Practice, 3rd edition	American Association of State Highway & Transportation Officials, Washington, D.C
AASHTO GDPS-4-M	Guide for Design of Pavement Structures, 4th edition, 1993 with 1998 supplement	American Association of State Highway & Transportation Officials, Washington, D.C

References' Key Chapters

General

The following is a compilation of crucial chapters and their corresponding key points for each reference, it is recommended to use this list as a review aid to ensure good coverage of the material. However, it is advisable to thoroughly examine all references and their chapters to adequately prepare for the exam.

Once you have thoroughly studied, practiced, and became proficient in the chapters listed below, and have ensured that you are entirely confident with the material covered and have practiced few examples as well, whether from this book, or examples that you can re-write yourself from this book or elsewhere, you may proceed to review other chapters in those references at your own pace.

Finally, it is important to note that the tables below may not correspond with the version of the reference material you are using, as recent versions may have altered chapters or sections' names or numbers or titles. An equation may have been added or omitted for instance. Therefore, it is important to be mindful of the version of the reference you are studying from as well as the version that will be used for the exam during your examination intake.

Chapters per Reference

Code	TITLE	No. of Chapters
HCM	Highway Capacity Manual, 6th edition and 7th edition	38 Chapters in 4 volumes
MUTCD	Manual on Uniform Traffic Control Devices for Streets and Highways, 2009	9 Parts
HSM-1	Highway Safety Manual, 1st edition, 2010	17 Chapter in 4 Parts
FHWA HIF-12-026	Hydraulic Design of Highway Culverts	6
AASHTO GDHS-7	A Policy on Geometric Design of Highways and Streets 7th edition, 2018	10
AASHTO GPF-2	Guide for the Planning, Design, and Operation of Pedestrian Facilities, 2nd edition, 2021	4
AASHTO RSDG-4	Roadside Design Guide, 4th edition, 2011	12
AASHTO MEPDG-3	Mechanistic-Empirical Pavement Design Guide, 3rd edition	13
AASHTO GDPS-4-M	Guide for Design of Pavement Structures, 4th edition, 1993 with 1998 supplement	4 Parts

HCM Manual Volume 1: Concepts (*)

This Volume consists of nine chapters in its 6th edition. A high-level summary is provided for some of the key chapters as noted below.

CHAPTER	TITLE & DESCRIPTION
1	**HCM User's Guide**
2	**Applications**
3	**Model Characteristics** This chapter covers various modes of transportation: motorists, trucks, pedestrians, bicycles, and transit. The chapter defines the K-factor (proportion to AADT that occurs during peak hour) and the D-factor (proportion of traffic moving in the peak direction) with examples and important equations such as: $$DDHV = AADT \times K \times D$$ The chapter provides different characteristics for vehicle classes such as length and power along with typical percentage of trucks on freeways and highways. The chapter continues talking about various modes of transport and their interactions with each other from a high level and literature perspective.
4	**Traffic Operations and Capacity Concerns** This chapter contains equations for speed, flow rates, Peak Hour Factor PHF, density, headway, spacing, pedestrians flow rates and their density, etc.
5	**Quality and Level-of-Service Concepts** This chapter provides an overview for Level of Service LOS. Every LOS is explained and defined in further details in their respective chapters.
6	**HCM and Alternative Analysis Tools** This chapter provides a high-level explanation on the available tools for analysis and those can be categorized at the operation's level, application's level, and the planning level. In a nutshell, this chapter explains how and why those tools were developed along with their use.
7	**Interpreting HCM and Alternative Tool Results** This chapter covers uncertainty and reliability of the results generated by the HCM tools.
8	**HCM Primer** This chapter explains some of the traffic engineering concepts. The last section of this chapter lists important companion documents most of which, if not all, are referred to in this book.
9	**Glossary and Symbols**

HCM Manual Volume 2: Uninterrupted Flow (*)

This Volume consists of six chapters its 6th edition. A high-level summary is provided for some of the key chapters as noted below.

CHAPTER	TITLE & DESCRIPTION
10	**Freeway Facilities Core Methodology** Apart from the core concepts this chapter provides, some of the important highlights are as follows: ○ Definition of segments and sections. ○ The calculation of segment(s) average density – Equation 10-1 ○ Level of Service LOS for freeway facilities Exhibit 10-6 ○ Data sources, such as Base Free Flow Speed BFFS and others per Exhibit 10-7 ○ Equations 10-4 to 10-6 for capacity and speed adjustment calculations CAF and SAF.
11	**Freeway Reliability Analysis** This chapter discusses reliability analysis and how travel time represented by speed and capacity is affected by various incidents/events. Few important exhibits are: ○ Exhibit 11-20 provides Capacity Adjustment Factors CAF due to weather conditions. ○ Exhibit 11-21 provides Speed Adjustment Factors SAF due to weather conditions. ○ Exhibit 11-23 provides capacity adjustment factors CAF due to incidents and closed lanes events.
12	**Basic Freeway and Multilane Highway Segments** This chapter defines key characteristics for freeways and highways. Exhibit 12-4 for instance compares capacities for both at different Free Flow Speeds. Equation 12-1 is an important equation that calculates mean speed based on capacity and breakpoint BP. Exhibit 12-6 is an important exhibit that contains various equations that define density, capacity and FFS. FFS is explained and calculated using Equations 12-2 and 12-3 for both freeways and highways respectively. LOS then determined towards the end of the chapter. Managed lane segments such as High Occupancy Vehicles lanes or Toll lanes are discussed in this chapter. Section 4 of this chapter provides methods for calculating adjusted speeds for those lanes – check Equations 12-12 to 12-19 along with Exhibit 12-30 that defines parameters based on different barriers.
13	**Freeway Weaving Segments** This chapter sets the maximum length that defines when a segment can be considered either a weaving segment or a merge/diverge segment – check Exhibit 13-11. It then talks about the various

configurations of one-sided and two-sided weaving segments.

The chapter provides equations that help determine weaving section capacity, average speed for weaving and non-weaving vehicles, weaving intensity, rate of lane change and density. It then defines Level of Service LOS assignments based on all of this.

An interesting application is the capacity assessment of a weaving segment that consists of a ramp on one side and a Managed Lane on the other side – check Equation 13-24 for further details.

14	**Freeway Merge and Diverge Segments**

The chapter provides methods for computing on-ramp and off-ramp density, capacity, and Level of Service. Exhibit 14-6 presents a flowchart with a step-by-step process for this purpose.

It is important to distinguish between on-ramp (merge) methods and off-ramp (diverge) methods, also it is important to understand how they influence each other if a freeway contains more than one – check L_{EQ} in Equations 14-7, 14-12 and 14-13. Also, attention should be made for equations used for four, six and eight lanes' freeways.

15	**Two-Lane Highways**

This chapter starts with defining "Two-Lane Highways" as those highways that have one lane in each direction along with their function, use and importance.

Particular attention is to be paid for this chapter as those types of highways have some certain methods for analysis.

The chapter dives into three classes of those highways and understanding these classes is of an utmost importance. For instance, in class I two-lane highways motorists are expected to travel at high speeds, class II highways serve as scenic or recreational routes, and class III are those highways that pass through small towns.

Level of Service is defined for each class in Exhibit 15-3 and in Exhibit 15-4 for bicycles.

Exhibit 15-6 presents an important flowchart which helps determine which methodology to be used for each class of highway to determine their relevant.

In a similar fashion, Exhibit 15-37 provides a similar flowchart and methodology but for determining bicycle LOS.

HCM Manual Volume 3: Interrupted Flow (*)

This Volume consists of nine chapters in its 6th edition. A high-level summary is provided for some of the key chapters as noted below.

CHAPTER	TITLE & DESCRIPTION
16	**Urban Street Facilities** This chapter sets the pace for Chapter 18 on urban street segments. Speaking of which, most of the important Exhibits can be picked up in Chapter 18. Chapter 16 can be referred to occasionally if a concept on an urban facility was rather needed. Some interesting Exhibits are: 16-2 which defines the isolating distance that allows a section or a segment to operate isolated from the rest of the urban boundary. Exhibits 16-3, 16-4 and 16-5 define the LOS criteria for vehicles, pedestrians, and bicycles respectively (all repeated in Exhibits 18-1, 18-2 and 18-3). Equations 16-3, 16-4, 16-5 and 16-6 calculate the travel speed, stop rate, pedestrian spacing and speed for the facility in its entirety based on the sum of its segments L_i. Equations 16-9 and 16-12 do the same for bicycles and transit speed.
17	**Urban Street Reliability and ATDM** This chapter measures and analyzes the reliability of urban streets using numerous factors and variables. The methodology of applying various scenarios to calculate reliability is presented.
18	**Urban Street Segments** This chapter evaluates the operation of urban street travel modes for vehicles, pedestrians, bicycles, and transit. Some of the important sections, exhibits and equations are listed below: Exhibit 18-1 defines Level of Service LOS for motorized vehicles in urban streets. Exhibit 18-2 does the same for pedestrians, Exhibit 18-3 for bicycles and transits. Exhibit 18-5 describes the required data and how to secure it to move ahead with the rest of the vehicles' methodology described in this chapter. Exhibit 18-8 presents the flowchart for motorized vehicles and it consists of *ten* steps. Some of the important steps are the ones relevant to speed, delay, signal, and signals spacing, vehicles proximity, proportion arriving when green, travel speed, stop rate and LOS. Exhibit 18-16 describes the required data and how to secure it to work on the pedestrians' methodology described in this chapter. Exhibit 18-17 presents the flowchart for pedestrians' methodology. This exhibit consists of *ten* steps as well. Some of the important steps are the ones relevant to determining Free Flow Speed FFS for pedestrians, average pedestrians' spacing, delay at

intersections and travel speed. Roadway crossing difficulty factor is an interesting subject worth considering as well.

Exhibit 18-23 presents the required methodology to determine LOS for bicycles.

The transit section is worth noting as well. Some of the important highlights in this section are the methodology flowchart of Exhibit 18-26, the equations relevant to speed such as Equations 18-48, 49 and 50.

19 Signalized Intersections

This chapter provides methodologies for vehicles, pedestrians, and bicycles at signalized intersections.

Some of the important highlights of this chapter are not limited to the following concepts, equations, and exhibits:
- o Phase duration and lost time Equations 19-1 and 19-2.
- o Level of Service LOS criteria Exhibit 19-8 for vehicles and 19-9 for pedestrians.
- o Sources of data per Exhibit 19-11. For instance, this exhibit defines base saturation flow rates for different populations.
- o Platooning Ratio in Exhibit 19-13 which helps define effective green time per Equation 19-5.
- o Determining adjusted saturation flow rates per Equation 19-8 along with the adjustment factors that follow. It is to be noted that some adjustment factors can only be determined using Chapter 31 – for instance for right turns, left turns, shared or exclusive lanes – there are few examples in this guide which cover those.
- o Determine capacity and volume to capacity ratio per Equations 19-16, 19-17, and 19-31. The same equations can also be used to determine the effective green time in specific contexts.
- o Determine delay per Equations 19-18 to 19-25 and understanding the queuing diagram or Queue Accumulation Polygon QAP per Exhibit 19-24.
- o Calculating pedestrian available time-space per Equation 19-51.
- o Compute the holding area waiting time for various conditions along with the circulation time. Calculate pedestrian service time and crossing occupancy – summing all the above using equations up to Equation 19-66.
- o Understanding pedestrian delay and how this affects LOS.
- o Compute bicycle delay and using this to determine LOS for the bike lane.

20 TWSC Intersections

This chapter explains how to compute lane capacities, movement control delay and intersection control delay for TWSCI Two-Way Stop-Controlled Intersections. Exhibit 20-6 presents a flow chart that takes you through the entire process.

Some of the other interesting highlights are the means and methods, and equations, used to understand conflicting traffic for right and left turns and other movements – check Equations 20-2 to 20-29. Also measuring headway per Equation 20-30 and follow-up

	headway per Equation 20-31, followed by calculating potential capacities per Equation 20-32 for no signals and up to Equation 20-35 for other conditions. The rest of the chapter takes you through the process of adjusting potential capacities for those lanes.		(2) two-lane approaches with single-lane R/A. (3) one-lane approach with two-lane R/A. (4 & 5) two-lane entry conflicted by two-lane R/A. (6 & 7) yielding bypasses opposed by one or two exit lanes.
21	**AWSC Intersections** This chapter presents methods for intersections that are All-Way Stop-Controlled. The methods presented in this chapter are iterative in nature. Unlike the R/A and the TWSC intersections' chapters, this chapter deals with lanes' saturations (Equations 21-1 to 21-5) and probabilities of conflicts (Exhibit 21-5 and Equations 21-6 to 21-10)	23	**Ramp Terminals and Alternative Intersections** This chapter goes through, explains, and defines all possible interchanges, their modifications, their use and some of their advantages and disadvantages. It then provides methods to calculate some of the adjustment factors required, for instance, for lane utilization, volume, saturation flow, etc. Check Exhibit 23-4 for a summary comparison on this. This chapter however does not provide geometric design advise for those interchanges. If required, other references such as AASHTO's Green Book should be consulted.
22	**Roundabouts** This chapter takes you through the process of determining the Level of Service for each of the roundabout R/A legs and approaches and to compute the average control delay along with the 95th percentile queues – check Exhibit 22-10 for a full methodology flowchart. Some of the interesting highlights and equations are the ones related to computing capacities for those lanes based on R/As circulating and conflicting traffic – i.e., conflicting to the lanes under consideration. For instance, Equations 22-1 to 22-7 provides calculating capacities for: (1) single-lane R/A with single-lane approaches.	24	**Off-Street Pedestrian and Bicycle Facilities** This chapter provides methodologies for calculating Level of Service for off-road facilities mainly for pedestrians and bicycles. Exhibit 24-7 provides a flowchart and a step-by-step method for exclusive off-street pedestrian facilities and Exhibit 24-11 does the same for bicycle lanes.

HCM Manual Volume 4: Applications (*)

This Volume's title is self-explanatory, and it consists of 14 chapters in its 6th or 7th edition. This volume is available online on the Transportation and Research Board TRB website.

As implied from the title of this volume, those chapters consist of detailed examples and applications which go along with volumes 1 to 3. The names of the chapters below are self-explanatory and sufficient to understand what examples they include.

I advise you to go over those examples at least at a high-level to understand the application of the concepts explained above.

CHAPTER	TITLE & DESCRIPTION
25	Freeway Facilities Supplemental
26	Freeway and Highway Segments: Supplemental
27	Freeway Weaving: Supplemental
28	Freeway Merges and Diverges: Supplemental
29	Urban Street Facilities: Supplemental
30	Urban Street Segments: Supplemental
31	Signalized Intersections: Supplemental
32	Stop Controlled Intersections: Supplemental
33	Roundabouts: Supplemental
34	Interchange Ramp Terminals: Supplemental
35	Pedestrians and Bicycles: Supplemental
36	Concepts: Supplemental
37	ATDM: Supplemental
38	Network Analysis

(*) Navigating the HCM Manual

This manual is quite lengthy and contains 38 chapters in total with around 2,300 pages. Our goal was to cover as much as we could, including at least 1 or 2 practice questions for each key "exam relevant" chapter or knowledge area. There's a lot to learn from this manual, and we've structured the questions strategically. This way, you will become familiar with where to find important parts, equations, and exhibits. This familiarity will be helpful during the actual exam, whether there was an example that covers a specific concept in this book or not.

MUTCD Manual

This manual consists of nine parts. A very high-level summary is provided for some of the key chapters as noted below.

PART	TITLE & DESCRIPTION
1	**General** This part consists of one chapter only and it talks about the design, placement, maintenance and the responsibility and authority when it comes to traffic control devices.
2	**Signs** This part consists of 14 chapters and those chapters cover everything that is relevant to signs: regulatory signs, barcodes and gates, warning signs and object markers, guide signs, toll roads and managed lanes and general information signs, etc. Finally, it covers the emergency management singing.
3	**Markings** This part consists of 10 chapters and those chapters cover everything that is relevant to markings: pavement and curb, roundabouts, preferential lanes, toll plazas, delineators, colored pavements, channelized devices, islands and finally rumble strips.
4	**Highway Traffic Signals** This part consists of 14 chapters and those chapters cover everything that is relevant to traffic signals: warrants (this is an important chapter that talks about studies which cover the need to install traffic signals to start with), traffic signal features, pedestrians control features and hybrid beacons, emergency vehicle access, traffic signals for one lane and two lane facilities and movable bridges, signals for plaza and flashing beacons, lane-use control signals and roadway lights.
5	**Traffic Control Devices for Low-Volume Roads** This part covers the need for signs, regulatory signs, markings and controls for low volume roads and school areas.
6	**Temporary Traffic Control** This part consists of nine chapters, and they cover signs, markings, and other temporary traffic control mechanisms for work zones and temporary closures and incident management. Also, particular attention should be paid to Typical Applications in Chapter 6H.
7	**Traffic Control for School Areas**
8	**Traffic Control for Railroads and Light Rail Transit Grade Crossings**
9	**Traffic Control for Bicycle Facilities**

HSM Manual

The HSM manual consists of four parts and 17 chapters, those chapters are spread across the four parts as follows:

PART	TITLE & DESCRIPTION
A	**Introduction, Human Factors and Fundamentals** This Part consists of three chapters as follows: Chapter 1 Introduction and overview Chapter 2 Human factors Chapter 3 Fundamentals Those chapters provide information that are mainly relevant to drivers and fundamentals of roads, crashes, and crash analysis. For instance, information such as vision, viewing distance, perception time or drivers' processing information and their attention can be found in Chapter 2. While information about crash rates, incidents, CMF Crash Modification Factors' definitions – more details on CMFs can be found in Part D.
B	**Road Safety Management Process** This Part consists of six chapters as follows: Chapter 4 Network Screening Chapter 5 Diagnosis Chapter 6 Select Countermeasures Chapter 7 Economic Appraisal Chapter 8 Prioritize Projects Chapter 9 Safety Effectiveness Evaluation Chapter 4 contain information that helps rank sites. The chapter consists of means and methods for how to choose from various performance measures, performance measures such as: average crash frequency, crash rates, EPDO, RSI, etc. It also provides screening methods such as: the sliding window method, the simple ranking method and the peak search method. It finally offers overall evaluation methods. Chapter 5 provides methods that help evaluate crash data and identify crash patterns. It also includes information of how to draw and analyze a collision diagram. Chapter 6 provides methods on how to identify factors contributing to crashes and helps select the best countermeasures for that. Chapter 7 provides means and methods for calculating benefit cost ratios along with methods of conducting economic appraisals. Chapter 8 evaluates economic justifiability for improvements. Chapter 9 evaluates effectiveness for countermeasures for one or multiple sites.
C	**Predictive Methods** This Part consists of three chapters, and they cover predictive methods for estimating expected average crash frequencies including severity and collision type as covered in the following: Chapter 10 Rural Two Lane Roads Chapter 11 Rural Multilane Highways

		Chapter 12 Urban and Suburban Arterials
	D	**Crash Modification Factors**
		This Part consists of five chapters, and they provide information on countermeasures, along with means and methods supported by tables and diagrams showing how to calculate Crash Modification Factors CMFs for: road segments, intersection, interchange, and other special facilities. Chapters covered here are as follows: Chapter 13 Roadway Segments Chapter 14 Intersections Chapter 15 Interchanges Chapter 16 Special Facilities and Geometric Situations Chapter 17 Road Networks

AASHTO's Green Book GDHS-7

This book consists of ten chapters in its 7th edition. A high-level summary for some of the key chapters is provided below. We advise you however to go through all the chapters and be ready to use the search function during the exam when in doubt.

CHAPTER	TITLE & DESCRIPTION
1	**New Framework For Geometric Design**
2	**Design Controls and Criteria**
3	**Elements of Design** This chapter defines sight distances, such as the Decision Sight Distance DSD, the Stopping Sight Distance SSD, and the Passing Sight Distance PSD, and how those can be used during the design process of vertical and horizontal curves. The chapter defines horizontal alignments, superelevations, and various methos for attaining superelevation transitions. The chapter also presents design methods and tables that can be used for calculating curve length and minimum radius, taking into account design speed and other factors. Design criteria for compound curves and spiral curves are discussed in this chapter as well. Off tracking and lane widening are important aspects covered in this chapter which require careful considerations. Moreover, vertical alignments criteria are detailed in this chapter with minimum and maximum allowable grades, calculating clearances and vertical (sag or crest) curve lengths. Escape ramps, and considerations when designing horizontal and vertical curves combined.
4	**Cross-Section Elements**
5	**Local Roads and Streets**
6	**Collector Roads and Streets**
7	**Arterial Roads and Streets**
8	**Freeways** Those five chapters provide important information that can be included in a question in the exam in the form of a requirement for a certain width of a cross-section, shoulder, lane or a median, a certain slope or a grade, or a requirement for a certain speed or superelevation for any of the above types of roadways. Chapter 7 also contains methods for attaining superelevation for divided roadways.
9	**Elements of Design** The most important information that you will encounter in this chapter is intersection sight distances found in Section 9.5. There is other important information that you will probably need to look into during the exam especially when it comes to curbs' sizes and areas, road channelization, track crossings and sight distances, auxiliary lanes, deceleration distance and storage length, medians and roundabout design and layout.

10	**Grade Separations & Interchanges**
	This chapter introduces various types of interchanges, roundabouts and layouts that can be referred to in case of a certain layout arises during the exam.
	Some of the important references, tables and figures of this chapter are the following:
	o Maximum grade design per Table 10-2. o Maximum length of taper per Table 10-3. o Acceleration lane length per Table 10-4 along with the relevant figures. o Grade adjustment factors for acceleration and deceleration lanes per Table 10-5. o Minimum deceleration lane length per Table 10-6 along with the relevant figures.
	Grade separation per Section 10.8.7 is worth exploring along with the section's relevant figures.

The Roadside Design Guide RSDG-4

This book consists of 12 chapters in its 4th edition. A high-level summary for some of the key chapters is provided below. We advise you however to go through all the chapters and be ready to use the search function during the exam when in doubt.

CHAPTER	TITLE & DESCRIPTION
1	**An Introductory to Roadside Safety**
2	**Economic Evaluation of Roadside Safety**
3	**Roadside Topography and Drainage Features** This chapter defines the clear-zone concept. It explains, by use of tables, methods for calculating clear zones and runout areas for recoverable ≤ 1:3 and nonrecoverable slopes > 1:3. The chapter covers the design of drainage cross sections as well. There are several examples and applications at the end of this chapter that you can go through to strengthen your understanding.
4	**Sign, Signal, and Luminaire Supports, Utility Poles, Trees and Similar Roadside Features**
5	**Roadside Barriers** This chapter presents numerous barrier types. Although no questions were posed in this guide on types, it is advisable to check them out.
	The chapter talks about selection criteria for those barriers, their maintenance, and their design (i.e., offset – Table 5-7, flare, and flare rate – Table 5-9, runout length – Table 5-10). There are few numerical examples on barrier length and flares at the end of this chapter that you can refer to.
6	**Median Barriers** Similar to Chapter 5 Roadside Barriers, this Chapter discusses, the different types of barriers that are used in medians, their applications and their maintenance.
7	**Bridge Railing and Transitions**
8	**End Treatments** This chapter discusses numerous end treatments that are presented in tables and in photos. One of the critical sections is section 8.4.3 which provides equations as to how the conservation of moment principle woks along with sand filled barrels.
9	**Traffic Barriers, Traffic Controls Devices, and Other Safety Features for Work Zones**
10	**Roadside Safety in Urban or Restricted Environments**
11	**Erecting Mailboxes on Streets and Highways**
12	**Roadside Safety on Low Volume Roads and Streets**

Hydraulic Design of Highway Culverts FHWA HIF-12-026

The FHWA HIF-12-026 consists of six chapters and three appendices. Most of the critical information, tables and graphs that are needed in this manual are also included in the *NCEES Handbook version 2.0* Section 6.4.10 Stormwater Collection and Drainage. FHWA HIF-12-026, being the official manual has more information.

Some of the important chapters and appendices that could be handy during working out an example, or during a design process, are the following – we encourage you however to go through all chapters:

Chapter 3 – Culvert Hydraulic Design
Understand the types of controls such as inlet control – Figure 3.1 and the relevant sections, and outlet control, Figure 3.7 and the relevant sections. The factors that influence those controls and the equations needed to calculate energy heads and head losses – the same equations are also included in the NCEES Handbook Section 6.4.10.2 Inlet and Outlet Control.

Appendix A – Inlet Control Equations
Contains equations for use for the inlet control for different shapes and various submerging conditions (for example headwater depth HW above a certain inlet section, etc.) – tables with inlet control constants are included here as well.

Appendix B – Hydraulic Resistance of Culvert Barrels

Appendix C – Design Charts, Tables, and Forms
Headwater depth for different types of culverts' material, shapes, wingwalls and orientations for inlet and outlet control.

Guide for the Planning, Design and Operation of Pedestrian Facilities GFP-2

The AASHTO GFP-2, which is the edition in use for the exam and for the year 2021, consists of four chapters. You will notice that most of the important information included in this guide is also available in the AASHTO's Green book, the MUTCD, the HCM Manual which all have been referred to in this book along with examples for pedestrian facilities' planning or design.

The four chapters are as follows:

Chapter 1: introduction

Chapter 2: Planning for Pedestrians

Chapter 3 Pedestrian Facility Design

Chapter 4: Pedestrian Facility Operations, Maintenance and Construction

We will not go into details or provide a summary for any of the above chapters as they are self-explanatory. My advice is to go through this easy read guide and arm yourself with the search function during the exam when faced with relevant questions.

Mechanistic-Empirical Pavement Design Guide MEPDG-3

This guide consists of 13 chapters. It details pavement design methods that are based on empirical studies.

The below is a summary for some of the important aspects of design, soil testing, rehabilitation methods that you should be aware of, and you should be able to refer to during the exam.

The below is a selected summary for some of the important chapters you are advised however to familiarize yourself with all chapters of this manual prior to the exam.

CHAPTER	TITLE & DESCRIPTION
1	Introduction
2	Referenced Documents and Standards
3	Significance and Use of the MEPDG
4	Terminology and Definition of Terms
5	Performance Indicator Prediction Methodologies
6	General Project Information
7	Selecting Design Criteria and Reliability Level
8	Determining Site Conditions and Factors
9	Pavement Evaluation for Rehabilitation Design This chapter presents methods for conducting initial pavement assessment (Table 9-1) along with some of the assessment activities and field tests (Table 9-2 & 3). Also, a guideline is presented as to what type of data input needs to be collected (Table 9-4) for assessment purposes along with methods of measurement. Destructive tests are presented in Table 9-6. And most importantly, relationships and equations to obtain CBR, M_r, E, R-value, etc. are presented in Table 9-7.
10	Determination of Material Properties for New Paving Materials
11	Pavement Design Strategies This chapter presents strategies for improving problem soils (Figure 11-1), stabilizing subgrades and other layers along with methods of doing so and what material needs to be added for this purpose. Flexible design and rigid design strategies are covered here as well. In those strategies, discussions per layer are presented, such as treatment of different layers, joints, spacings, tying up with other layers or with an existing concrete shoulder, etc.
12	Rehabilitation Design Strategies
13	Interpretation and Analysis of the Trial Design

Guide for Design of Pavement Structures GDPS-4-M

This guide consists of four parts and each part has four to five chapters in it. The guide contains nine appendices as well, some of the important ones are:

- Appendix D for the calculation of ESAL,
- Appendix G with the treatment of swelling and frost heave.
- Appendix H and G provide a design example for flexible and rigid pavements respectively.

The below is a summary for some of the important parts of this guide. You are advised however to familiarize yourself with all parts/chapters and appendices prior to the exam.

PART	TITLE & DESCRIPTION
1	**Pavement Design and Management Principles** This part contains design equations for rigid and flexible pavements in its first chapter along with relationship equations for CBR, R-value and the resilience modulus M_r. This first chapter also has cross sections for drainage, and it defines and identifies permeability for aggregates in Table 1.1.
2	**Pavement Design Procedures for New Construction or Reconstruction** This part discusses methods for design and construction, and it provides various design charts (nomographs) that can be used to determine rigid and flexible pavement various layers thicknesses – this is provided in Chapter 3. Chapter 1 defines and calculates ESALs in conjunction with Appendix D. Chapter 4 repeats the above design methods but for low volume roads with maximum ESAL between 700k to 1 million.
3	**Pavement Design Procedures for Rehabilitation of Existing Pavements** This part presents various rehabilitation methods with and without overlay.
4	**Mechanistic-Empirical Design Procedures** This is a short part that covers few tests and better to refer to the new edition of the Mechanistic-Empirical Pavement Design Guide MEPDG-3 for more details on this area.

Your Feedback Matters – Make Sure You Share It With Others

Good day,

As you reach the final pages of this book, we would like to express our sincere gratitude for choosing it as your guide to aid you in your journey toward success in the PE exam. We have poured countless hours into meticulously crafting the questions and practice exams within these pages.

Your opinion matters greatly in helping others discover the value of this resource. If you found this book beneficial, kindly consider leaving your positive and honest feedback on the platform that you bought it from - like Amazon. Your words will not only acknowledge the hard work invested into producing this book but will also guide future readers in their quest for quality study materials.

Remember, your review is more than just feedback; it's a beacon for those seeking reliable resources. Your support can make a significant difference, ensuring that this book continues to assist aspiring professionals on their path to success.

Thank you for being a part of this journey, and we appreciate your commitment to sharing your experience with others.

Best wishes,

PE ESSENTIAL GUIDES

Made in the USA
Las Vegas, NV
04 July 2024

9186l567R00116